Revise A2 Biology for AQA Specification B

Graham Read and Ray Skwierczynski

Heinemann Educational Publishers,
Halley Court, Jordan Hill, Oxford OX2 8EJ
Part of Harcourt Education

Heinemann is the registered trademark of Harcourt Education Limited

First published 2002

ISBN 0 435 58313 1

05 04
10 9 8 7 6 5 4

Development editor Paddy Gannon

Edited by Margaret James

Designed and typeset by Saxon Graphics Ltd, Derby

Index compiled by Ann Hall

Printed and bound in UK by Thomson Litho Ltd

Tel: 01865 888058 www.heinemann.co.uk

Contents

Introduction – How to use this revision guide

This revision guide is for the AQA Biology A2 course. It is divided into modules to match Specification B. You may be taking a test at the end of each module, or you may take all of the tests at the end of the course. The content is exactly the same.

Each module begins with an **introduction**, which summarises the content. It also reminds you of the topics from your GCSE course which the module draws on.

The content of each module is presented in **blocks**, to help you divide up your study into manageable chunks. Each block is dealt with in several spreads. These do the following:

- summarise the **content**;
- indicate **points to note**;
- include **worked examples** of calculations;
- include **diagrams** of the sort you might need to reproduce in tests;
- provide **quick check** questions to help you test your understanding.

In the margin of each page are shaded boxes.

 **Tip boxes** contain information that adds to what is in the text, or points out common mistakes.

 **Synoptic tip boxes** point out the sort of links with other modules you should know – to answer synoptic questions.

At the end of each module, there are longer **end-of-module questions** similar in style to those you will encounter in tests. **Answers** to all questions are provided at the end of the book.

You need to understand the **scheme of assessment** for your course. This is summarised on page vii. At the end of the book, you will find some **exam tips** to help prepare you for the examinable component of the course.

AQA A2 Biology – Assessment

To get a full A level award for AQA Specification B you need an AS (forming 50% of an Advanced Level), followed by the A2 (for the other 50%).

The **AS** has three compulsory Assessment Units (Modules): Core Principles, Genes and Genetic Engineering, and Physiology and Transport. Coursework is also required and is dealt with by your school or college. To get an AS, you will have to take a $1\frac{1}{4}$ hour exam for each Module/Unit: Core Principles, Genes and Genetic Engineering, and Physiology and Transport. All the questions have to be answered in each exam. Most of the marks will be for remembering and being able to explain the information in each Module. There is no synoptic element in these exams (that comes at Advanced Level).

The **A2** has two compulsory Assessment Units – covered by this book. There is no synoptic element in **Unit 4, Energy, Control and Continuity**. There will be more questions requiring analysis of information than in the AS exams. The exam for **Unit 5, Environment** has a large synoptic content. Synoptic questions are explained in 'Exam Tips' but you will have to know important principles from all the other Units.

What you need to know is given in this book, at the level you need to know it. If you know more than that, it certainly will not harm you but it will not be needed to pass an exam.

At the end of each section of this book you will find quick check questions. A box in the text indicates when you should be able to attempt a particular question. At the end of each Module are exam-type questions for you to try. Answers to all of the questions are given at the end of the book.

✓ Quick Check 1

The table shows an outline of the AQA Specification B A2.

AQA Specification B – A2 Examination
Energy, Control and Continuity – Assessment Unit 4 $1\frac{1}{2}$ **hour exam**   **15% of total A level mark**
Environment – Assessment Unit 5(a) $1\frac{1}{4}$ **hour exam**   **7.5% of total A level mark**   **(including 3.5% synoptic)**
Coursework – Assessment Unit 5(b) [Not covered in this book]   **7.5% of the A2**
One of Assessment Units 6, 7 or 8 [Not covered in this book] $2\frac{1}{4}$ **hour exam**   **20% of total A level mark**   **(including 14% synoptic)** **Section A on your chosen Unit** **Section B Applying Biological Principles**

Module 4: Energy, Control and Continuity

This module is broken down into 12 topics: Energy supply; Photosynthesis; Respiration; Survival and coordination; Homeostasis; Nervous coordination; Analysis and integration; Muscles; Inheritance; Variation; Selection and evolution; and Classification.

Energy supply, Photosynthesis and Respiration

- Photosynthesis is a process in which light energy is used to make organic molecules.
- Respiration is the process in which energy in organic molecules is made available for all the other processes in a cell or organism.
- ATP is synthesised in respiration and used as the immediate source of energy for biological processes.
- Photosynthesis in chloroplasts involves light-dependent and light-independent reactions.
- The light-dependent reaction produces reduced coenzyme NADP and ATP.
- The light-independent reaction involves the reaction of carbon dioxide with an acceptor molecule and a reduction reaction that produces a sugar, using reduced NADP and ATP.
- Respiration makes energy from organic molecules available as ATP.
- Glycolysis is the first stage – the oxidation of glucose to pyruvate in the cytoplasm.
- The Krebs cycle takes place in mitochondria – pyruvate is oxidised to produce reduced coenzyme NAD and FAD and waste carbon dioxide.
- The reduced coenzymes provide energy for the production of large amounts of ATP by oxidative phosphorylation.
- Oxygen is used as a terminal electron acceptor in oxidative phosphorylation.

Survival and coordination

- Organisms respond to changes in their environment – stimuli.
- Stimuli involve energy changes which are detected by receptors.
- Receptors convert the energy associated with a stimulus into a form the cell or organism can understand – usually nerve impulses in animals.
- A coordinator processes information from receptors and formulates a response.
- The response is produced by effectors – muscles or hormones.

Homeostasis

- Homeostasis is maintaining a constant internal environment.
- It involves responses to internal and external stimuli.

- Negative feedback prevents systems from fluctuating too far from their normal state.
- Mammals control body temperature, because of its effect on enzyme activity.
- They control blood sugar, because of its importance as a respiratory substrate for cells.
- Protein metabolism produces toxic urea which the kidneys remove from the blood and excrete in urine.
- The kidneys also regulate blood water potential.

Nervous coordination

- Receptors detect stimuli and generate nerve impulses.
- The eye has light receptor cells in the retina – rods and cones.
- The cornea and lens focus images onto the retina and the iris and pupil control the amount of light entering the eye.
- Cones are responsible for colour vision and high visual acuity.
- Rods are not colour sensitive and give low visual acuity but are more sensitive to low light intensities than cones.
- Nerve impulses are separate travelling action potentials carrying information along neurones.
- Information is carried by the frequency of nerve impulses.
- Neurones communicate information to each other at synapses using chemical transmitter substances.
- Drugs can affect transmission of information across synapses.
- Some neurones transmit information to muscle fibres at neuromuscular junctions.

Analysis and integration

- The brain coordinates responses to stimuli.
- Sensory areas receive and process information from receptors.
- Association areas interpret sensory input.
- Motor areas control effectors.
- Understanding and producing speech is very important to humans and this is reflected in the large areas in the brain devoted to these abilities.
- There is an autonomic nervous system, divided into sympathetic and parasympathetic components with opposite effects.
- Pupil diameter, tear production and control of emptying of the bladder are under autonomic control.

Muscles

- Muscles are effectors that usually work in antagonistic pairs.
- Muscle cells have a highly specialised structure to allow them to contract.
- They contract when actin and myosin protein filaments slide past each other.
- This requires energy which is supplied by ATP.

Inheritance

- Organisms inherit genes which exist in different forms, or alleles.
- The genotype of an organism is its genetic constitution – the genes it has and which alleles of those genes.
- Meiosis produces new combinations of alleles through independent assortment and crossing over.
- It also halves the number of chromosomes in the cells it produces.
- Patterns of inheritance of genes can be predicted if the genotypes of the parents are known.
- Monohybrid crosses involve alleles of a single gene.
- Dihybrid crosses involve alleles of two different genes.
- The sex of humans is determined by the X and Y sex chromosomes; women are XX and men are XY.
- Whatever allele of a gene is on the X chromosome of a man, it is expressed.
- This leads to sex-linkage, where some genetically determined characteristics are more common in men than in women.

Variation, Selection and evolution

- Variation is found in the phenotypes of individuals in a population of a species.
- Variation between individuals is discontinuous or continuous.
- Discontinuous is the result of genetic factors (mainly) and is qualitative.
- It is often due to the action of one (or two) genes.
- Continuous is the result of genetic and environmental factors and is quantitative.
- It is often controlled by many genes – polygenic.
- Variation in a population leads to differential survival and reproduction.
- The phenotypes of some individuals are better suited to their environment than others.
- They are more likely to survive and reproduce, thus passing on their combinations of alleles to the next generation.
- This natural selection can lead to changes in the frequencies of alleles in a population.
- This is evolution and may lead to speciation – the formation of new species.

Classification

- Classification is hierarchical.
- Classification can also be phylogenetic, reflecting the evolutionary history and links of species.
- Organisms belong to one of five kingdoms, each with its own distinguishing features.

Energy supply – ATP, oxidation and reduction

Energy supply

All **cells (and organisms) respire all the time**, **including plant cells**. In **respiration** the chemical (covalent) bonds between the atoms of respiratory substrates (biological molecules) are broken. This releases energy that can be used to make **ATP**. ATP contains **chemical energy** that can be used for biological processes.

- **Photosynthesis** in plants makes organic glucose molecules from inorganic carbon dioxide and water molecules.

- These **synthetic reactions** make larger, more complex molecules from smaller, simpler molecules.

- **Light energy** is absorbed by chlorophyll and provides the energy input needed for the synthetic reactions.

- **Respiration** in plant cells uses a lot of the glucose produced by photosynthesis.

- Animals get respiratory substrates by **feeding** and plants make them in **photosynthesis**.

Ⓢ Plants are producers in food chains. They can make all the different types of biological molecules found in cells from sugars made in photosynthesis. Other organisms get these molecules from plants by feeding.

✓ *Quick check 1*

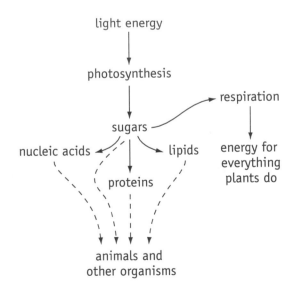

ATP

An ATP molecule is made by adding an inorganic **phosphate group** to **ADP**.

- This reaction needs an input of energy and is linked either to energy released when chemical bonds in glucose (or other carbohydrates, lipids or amino acids) are broken in respiration, or to the light-dependent reaction of photosynthesis.

- ATP can transfer a phosphate to another molecule – **phosphorylation**.

- The phosphorylated molecule is more reactive and this **lowers the activation energy** needed for reactions involving enzymes.

- ATP becomes ADP again when it loses its phosphate group.

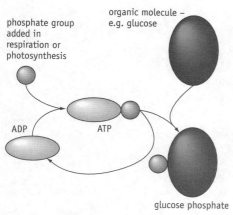

ATP and phosphorylation

Oxidation and reduction

- Many of the reactions in respiration and photosynthesis involve oxidation and reduction. There are **three types of oxidation reaction**:
  - removal of hydrogen;
  - loss of electrons;
  - combination with oxygen.

- Oxidation is linked to **reduction**, where oxygen is lost or hydrogen or electrons are gained by another atom, molecule or ion.

- The reduced substance is a source of chemical energy (**reducing power**).

(S) Enzymes also lower the activation energy of reactions.

▶ A lot of the biochemistry you have already learnt is simplified. Large molecules are actually made from smaller, phosphorylated molecules. For example, starch is made from glucose phosphate and proteins are made from phosphorylated compounds of amino acids.

✓ Quick check 2

▶ Make sure you remember that the removal of hydrogen from a molecule is an oxidation!

? ## Quick check questions

1 Students often write that animals respire but plants photosynthesise. Explain what is wrong with that statement.

2 Explain why ATP is necessary for reactions in cells to take place.

Photosynthesis

Plants absorb light energy and change it to chemical energy in bonds holding organic molecules together. Photosynthesis takes place in two stages, the light-dependent reaction and the light-independent reaction.

ⓈPhotosynthesis is the route by which energy enters food chains.

Light-dependent reaction

The light-dependent reaction uses light energy to make some ATP and a reduced coenzyme.

- Light energy is absorbed by electrons in **chlorophyll.**
- Some **excited electrons** gain enough energy to **leave chlorophyll.**
- These electrons are replaced from the **photolysis of water** – water molecules break down, releasing **H⁺ ions**, **electrons** and **oxygen.**
- Oxygen is lost as a waste product.

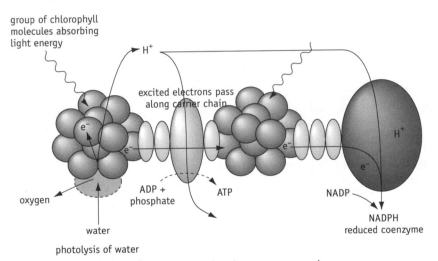

Light-dependent reaction in granum membrane

- The excited electrons **reduce** the first of a chain of **electron carriers – absorbed light energy has been converted into chemical energy (reducing power).**
- Electrons lose energy as they pass along the carrier chain and this is linked to making ATP by the addition of phosphate to ADP.
- At the end of the **carrier chain**, electrons **reduce coenzyme NADP to form NADPH.** (This is where electrons are reunited with the H⁺ from water.)
- **The reduced coenzyme is a reducing agent; a source of reducing power which is used in the light-independent stage of photosynthesis.**

A coenzyme is a non-protein which has to be present for an enzyme to work. The reduced coenzyme produced here becomes a source of hydrogen for an enzyme that adds hydrogen to a molecule.

Chlorophyll and the proteins of the electron carrier chain are located in the membranes of the **grana** of the chloroplast.

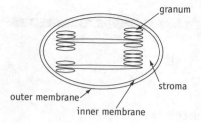

Chloroplast structure

Make sure you know where the two stages of photosynthesis take place in the chloroplast.

✓ Quick check 1, 2

Light-independent reaction

ATP and **NADPH from the light-dependent stage** are used in the **stroma** in the **light-independent** stage. Carbon dioxide (hydrogencarbonate ions) diffuses into the stroma along a concentration gradient.

- Carbon dioxide reacts with a five-carbon sugar, **ribulose bisphosphate**.
- This produces two, three-carbon molecules of **glycerate 3-phosphate**.
- This is converted to a **sugar** (**carbohydrate**) in a **reduction reaction**.
- The hydrogen for this comes from reduced NADP (**NADPH**).
- To lower the activation energy for the reactions, **ATP** is used to phosphorylate the reactants.
- Some of the sugars produced are used by the plant, the rest are used to **regenerate ribulose bisphosphate** (Calvin cycle).

Ⓢ Glycerate 3-phosphate is a dissociated acid that is reduced to a sugar by NADPH.

Ⓢ This stage relies on enzymes and its rate is affected by factors that affect enzymes, e.g. temperature.

Don't try to learn any more details of the biochemistry of photosynthesis. What you need to know is here!

✓s Quick check 3, 4

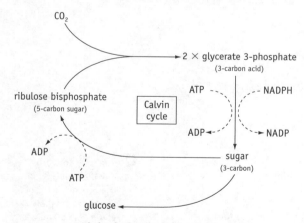

Light-independent reaction

? Quick check questions

1 Explain why water is needed for photosynthesis.

2 Explain how light energy is used in photosynthesis.

3 Radioactive carbon dioxide was given to a plant. Suggest the order in which molecules in the chloroplasts would become radioactive.

4 Explain why the light-independent stage of photosynthesis stops very quickly if a plant is put into the dark.

Respiration

The breaking of chemical (covalent) bonds in glucose (or other carbohydrates, lipids or amino acids) releases energy. Some of this energy is used to make ATP and the rest appears as heat. Glucose is broken down in a series of reactions involving **oxidations**, linked to **reduction** of **coenzymes**. Reduced coenzymes are the source of the energy (**reducing power**) used to make ATP.

Glycolysis

This is the first stage of respiration and takes place in the **cytoplasm. Six-carbon** glucose is oxidised into **two, three-carbon** molecules of **pyruvate**.

- Glucose is phosphorylated with two phosphate groups from two ATP.
- Glucose is **oxidised** by **removing hydrogen** and **ATP is produced**.
- Hydrogen is accepted by a **coenzyme**, **NAD**, forming **reduced NAD**.
- This oxidation of glucose produces two molecules of **pyruvate**.
- There is a **net gain of two ATP** (four are made – two are used at the start).

Glycolysis does not need oxygen. It is always the first stage of respiration. In anaerobic respiration it is the only stage.

✓ *Quick check 1*

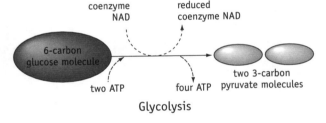

Glycolysis

Krebs cycle

In aerobic respiration, pyruvate from glycolysis enters the Krebs cycle in the matrix of the **mitochondrion**.

This **produces reduced coenzymes NAD and FAD**.

Don't try to learn any more details of the biochemistry of respiration. What you need to know is here!

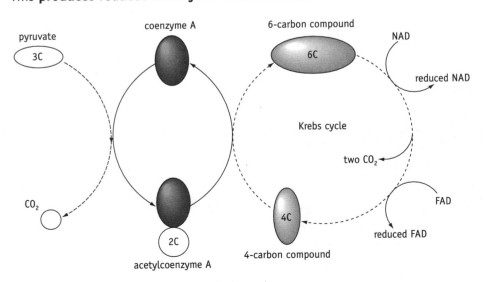

Krebs cycle

- **Pyruvate** reacts with **coenzyme A**, producing **acetylcoenzyme A** (carrying **two carbon atoms** from pyruvate) and releasing carbon dioxide.
- Acetylcoenzyme A reacts with a four-carbon compound, forming a **six-carbon compound**.
- **This is oxidised by removing hydrogen.**
- Hydrogen is accepted by coenzymes, to give **reduced coenzymes NAD and FAD** ($NADH_2$ and $FADH_2$) – the **main products of the Krebs cycle**.
- Carbon dioxide is released as a waste product and the four-carbon molecule that reacts with acetylcoenzyme A is reformed.
- A small amount of ATP is made by the Krebs cycle.

> ◖ Make sure that you answer the question set! **Don't** be tempted to write everything you have memorised about respiration – especially details of biochemistry not in the specification!

> ✓ *Quick check 2*

Oxidative phosphorylation

This is associated with cristae forming the inner mitochondrial membrane and produces almost all the ATP in aerobic respiration.

- Reduced NAD and FAD from glycolysis and the Krebs cycle reduce the first protein in an **electron transport chain**.
- The hydrogen from the coenzymes gives H^+ and an **electron**.
- The **electron loses some energy** as it passes down the transport chain.
- Some of this energy is used to **phosphorylate ADP to ATP**.
- This is the **oxidative phosphorylation** of ADP to make ATP.
- Electrons coming off the electron transport chain are accepted by **oxygen** and joined by H^+ to make water (another waste product of respiration).

> Ⓢ The heat released in respiration is energy which is lost to the environment from organisms and food chains.

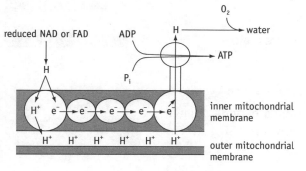

Oxidative phosphorylation

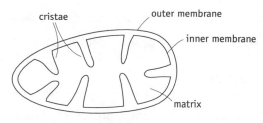

Structure of mitochondrion

> ✓ *Quick check 3, 4*

? *Quick check questions*

1 Describe the oxidation of glucose in glycolysis.

2 Explain how the Krebs cycle is linked to ATP production.

3 Explain how ATP is produced by oxidative phosphorylation.

4 Cyanide is a poison that stops the reactions combining oxygen, electrons and H^+ to make water. Suggest how this action of cyanide kills.

Survival and coordination

Stimulus and response

Organisms increase their chances of survival by responding to changes in their environment (external or internal). For example, animals respond to food, potential mates and predators. Plants respond to light, water and mineral ions. Animals and plants respond to internal changes in water content.

- A **stimulus** is a change in the environment, involving an energy change.
- **Receptors** detect energy changes.
- For example, receptors in the eye (rod and cone cells) detect changes in light energy and produce **nerve impulses**, a form of electrical energy.
- Nerve impulses carry **information** in the **nervous system**, to the **coordinator** – which decides the appropriate response to the stimulus.
- In humans/mammals, the coordinator is the spinal cord or the brain.
- The coordinator sends nerve impulses to **effectors – muscles or glands**.
- **Muscles** contract, producing a movement **response**.
- Glands release **hormones**, causing a change in physiology as a **response**.
- A response is what an organism/animal does in reaction to a stimulus.

The pathway linking a stimulus to a response is shown below.

stimulus ⇒ receptor ⇒ coordinator ⇒ effector ⇒ response

Ⓢ The ability to produce appropriate responses to environmental stimuli is part of the adaptations of an organism for the niche it occupies in a community and ecosystem.

▶ Make sure you always look for these things when answering a question: stimulus ⇒ receptor ⇒ coordinator ⇒ effector ⇒ response.

✓ *Quick check 1*

Simple reflex

A reflex is an **automatic** response, **protecting** the organism from a harmful stimulus. The **hand withdrawal reflex** happens if you put your finger onto something which is too hot.

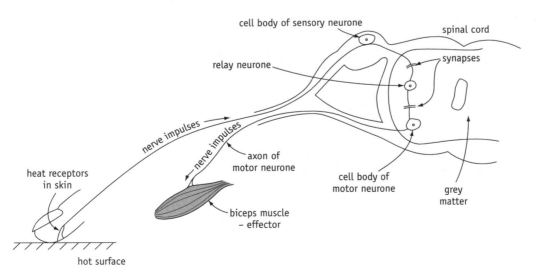

A simple reflex

- This involves **three neurones**.
- Heat **receptors** detect a harmful rise in skin temperature, the **stimulus**.
- Nerve impulses travel along a **sensory neurone** into the spinal cord (the **coordinator**), to a synapse with a **relay/association neurone**.
- A **synapse** is where information is transferred from one neurone to another.
- The relay neurone synapses with a **motor neurone**, which sends nerve impulses to the **effector**, the biceps muscle.
- This contracts, pulling the hand away from the heat – the **response**.

> Always use the terms used here. For example, nerve impulses are not 'messages'!

✓ *Quick check 2*

Hormones

Information can be transferred by **hormones**, which are chemical messengers.

- Hormones are secreted by **endocrine glands** directly **into the blood**.
- They affect the **physiology of target cells**.
- One hormone can target different types of cells in different parts of the body, but all the cells have receptors for the hormone.
- Hormones work in the body at very low concentrations.

Examples of hormonal regulation are given in the next section, Homeostasis.

✓ *Quick check 3*

? Quick check questions

1 When something frightening appears, our adrenal glands release the hormone adrenaline, which makes our body ready to run or fight. Suggest how the appearance of something frightening leads to the release of adrenaline.

2 Explain how our hand can be pulling away from a hot surface before we are even aware that we have touched it.

3 Adrenaline speeds up heart rate, dilates the air passages to the lungs and makes the pupils of the eye open wider. Suggest how the release of one hormone can lead to such a wide range of responses.

Regulation of blood sugar

Maintaining a constant internal environment is called **homeostasis**. An important function of homeostasis is to maintain **optimum conditions for enzyme activity**, in terms of factors like temperature, pH, substrate and product concentrations.

Negative feedback

Negative feedback is a control mechanism where movement away from the normal value of something produces a response that returns it to its normal value.

- For example, the product of a series of enzyme reactions can inhibit an enzyme in the series, reducing its own production.
- The result is that the concentration of the product rises and falls slightly, above and below a certain (optimum) concentration.

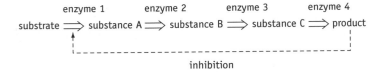

Regulation of blood sugar

The body maintains an optimum concentration of glucose in the blood (blood sugar), to supply cells with glucose for respiration. Two **hormones** control blood sugar concentrations, **insulin** and **glucagon** – secreted by the **pancreas**.

Blood glucose concentration rises above normal

- Blood sugar rises after a meal, when glucose is absorbed through the lining of the small intestine and into the blood.
- Cells in the **pancreas** respond by **secreting insulin** into the bloodstream.
- **Liver** cells have **specific membrane receptors** for insulin: proteins with a tertiary structure/3D shape and receptor site which only insulin fits into.
- Binding of insulin causes glucose channels (carrier proteins) to open, allowing more glucose into cells from the blood (by facilitated diffusion).
- This returns blood sugar levels towards normal.
- Increased uptake of glucose provides a higher substrate concentration for **enzymes** inside liver cells which convert glucose into **glycogen**, an insoluble storage carbohydrate.
- As blood glucose levels return to normal, insulin secretion is reduced. The secretion of insulin eventually causes a reduction in its secretion – **negative feedback**.

Ⓢ Meals with lots of glucose in them quickly raise blood sugar levels. Starch-rich meals have to be digested to glucose and this takes time – so blood sugar rises slowly. This is better for diabetics.

◖ See if you can explain the control of insulin release as an example of homeostasis and negative feedback.

✓ *Quick check 1, 2*

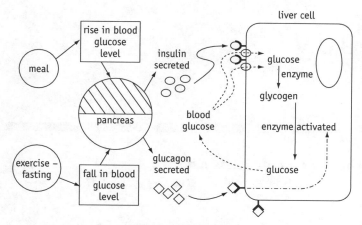

Control of blood sugar concentration

Blood glucose concentration falls below normal

- This can happen during exercise, or when someone hasn't eaten for some time.
- Cells in the pancreas respond by secreting **glucagon** into the blood.
- **Liver** cells have **specific membrane receptors** for glucagon.
- Binding leads to switching on of **phosphorylase enzymes** which break down glycogen into glucose.
- Glucose diffuses into the blood, returning blood glucose towards normal.
- The fall in blood glucose concentration causes a reduction in the release of glucagon – **negative feedback**.
- The **opposing actions of insulin and glucagon** keep blood sugar from fluctuating too much from the optimum concentration.

Don't write that insulin or glucagon react with, or act on, glucose or glycogen – they don't! These hormones affect liver cells!

✓ *Quick check 3*

❓ Quick check questions

1 Explain why insulin levels in the blood rise and fall at various times during the day.

2 Explain how insulin returns blood sugar levels to normal after a sugar-rich meal.

3 Explain the role of specific membrane receptors in the actions of insulin and glucagon.

Regulation of body temperature

To maintain a constant temperature mammals control heat loss. Heat is released from respiration. Heat can be lost by radiation, convection, or conduction to cold surfaces.

- **Thermoreceptors** in blood vessels in the **hypothalamus** (in the brain) detect rises or falls in blood temperature.
- Thermoreceptors in the **skin** detect changes in environmental temperature.

If blood or environmental temperature starts to rise

- Thermoreceptors send nerve impulses to a **heat loss centre** in the **hypothalamus**, the coordinator.
- This sends nerve impulses to effectors in the body, but mainly in the skin.
- The **responses** increase heat loss which lowers the temperature of the blood.

Vasodilation

- Relaxation of circular muscles occurs in walls of arterioles in the skin, allowing more blood to flow nearer the surface of the skin.

Sweating

- Nerve impulses cause sweat glands to release more sweat.
- This spreads over the skin surface and evaporates – getting the heat of vaporisation from the skin (and the blood flowing through it).

Flattening of hair

- Nerve impulses cause muscles at the base of each hair to relax, allowing the hair to lie flat.
- This reduces the thickness of the insulating air layer next to the skin and increases heat loss (reduces the heat holding).

Behaviour

- Humans can take off clothing, or find shade from the sun.

If blood or environmental temperature starts to fall

- Thermoreceptors send nerve impulses to the **heat gain centre** in the hypothalamus.
- This sends nerve impulses to effectors in the body, but mainly in the skin.
- The **responses** reduce heat loss (and sometimes cause more heat production).

effectors
vasodilation
sweating
flattening of hair
nerve impulses
increased
heat loss
heat
loss centre
thermoreceptors
in hypothalamus
heat
gain centre
blood and body
temperature rising
above normal
blood and body
temperature falls
towards normal

Regulation of body temperature

▶ Blood vessels – such as capillaries – do not move during vasodilation (or vasoconstriction)!

▶ All of these responses can increase heat loss by radiation, convection (warm air rising off the body), or conduction (if you are touching a cool surface).

✓ *Quick check 1*

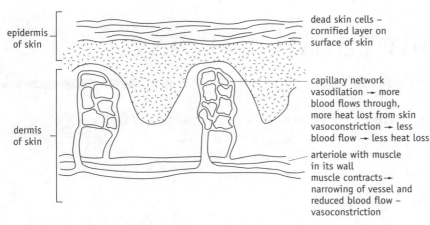

Vasodilation and vasoconstriction

⑤ During exercise, muscles need more energy from respiration. More heat is produced in muscles. This heat is carried away by the blood to the skin.

Vasoconstriction

- Nerve impulses cause contraction of circular muscles in walls of arterioles in the skin, allowing less blood to flow near the surface of the skin.

Sweating

- Less sweat is released.

Insulating effect of hair

- Muscles at the base of hairs contract and the hairs stand up.
- This increases the thickness of the insulating air layer next to the skin and reduces heat loss.

Shivering

- Nerve impulses cause muscles in the body to contract in little spasms, producing more heat from respiration.

Metabolic rate

- Nerve impulses cause the adrenal glands to secrete the **hormone adrenaline**.
- This causes an increase in the rate of respiration in body tissues and more heat production.

Negative feedback and inhibition

- When the heat loss centre is active, it sends nerve impulses that inhibit the heat gain centre.
- As heat loss increases, body temperature falls until it drops below normal.
- The heat gain centre then becomes active, inhibits the heat loss centre, and makes the body temperature rise again.
- This is control of body temperature by negative feedback.

⑤ If you have to write about the response of the body to a change in temperature, think about the sequence: stimulus ⇒ receptor ⇒ coordinator ⇒ effector ⇒ response.

✓ *Quick check 2, 3*

❓ *Quick check questions*

1. Explain how body temperature is kept constant during exercise.
2. Drinking an ice-cold drink can lower the temperature of blood flowing in vessels near to the stomach. Suggest how an ice-cold drink might cause a temporary reduction in heat loss from the skin.
3. People with hypothermia have a body temperature below normal. They become less and less active, and they can eventually become unconscious and die. Suggest why there is a progressive fall in body temperature.

Removal of metabolic waste

Waste products of metabolism are toxic/harmful if they accumulate. Animals cannot store amino acids or protein. Reactions in **liver cells** convert surplus amino acids into an organic acid, used in respiration, and toxic ammonia. Ammonia is used to make urea, which is released into the blood and excreted by the kidneys.

Don't learn lots of unnecessary details about the ornithine cycle!

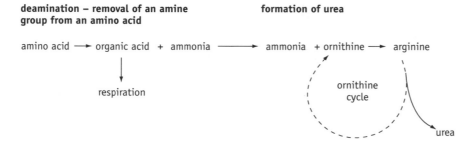

deamination – removal of an amine group from an amino acid

formation of urea

Production of urea

Urine production

✓ Quick check 1

The kidneys remove urea from the blood and concentrate it in urine. The kidney consists of vast numbers of tubules (nephrons). Different parts of each tubule are specialised for particular functions.

Ⓢ Make sure you know the general pattern of blood circulation. How does a molecule of urea get from the liver to the kidney in the blood?

Ultrafiltration

- Blood enters the kidney under high pressure through the renal artery.
- This branches into arterioles, then bundles of capillaries, the **glomeruli.**
- These capillaries have 'pores' between the endothelial cells, allowing substances to leave the blood without crossing cell membranes.
- There are also 'pores' between the cells (podocytes) of the **renal capsule**/Bowman's capsule surrounding the glomerulus.
- A lot of the blood plasma is forced through these 'pores' into the tubule by the **high blood pressure**, forming the **filtrate – ultrafiltration.**
- The pores act as a filter, **letting through** small substances like water and dissolved urea, glucose and mineral ions but **not** large blood cells and large blood proteins.

Ⓢ Blood consists of blood plasma and suspended blood cells. Plasma contains dissolved glucose, mineral ions, amino acids, urea and proteins.

✓ Quick check 2

Selective reabsorption of useful substances

Useful substances in the filtrate are **reabsorbed into the blood** along the tubule. The reabsorption of water is vital, to avoid dehydration.

Filtrate leaving the renal capsule enters the next part of the tubule.

Make sure you understand water potentials. Pure water has a water potential of 0. The more substances there are dissolved in it, the lower the water potential. There is a net flow of water by osmosis from a higher to a lower water potential.

Proximal convoluted tubule

- **Glucose, Na⁺ and K⁺** ions are reabsorbed from the filtrate by **active transport**, using **specific carrier proteins** in cell membranes of tubule cells.
- No glucose is left in the filtrate or excreted with urine.
- Active transport uses a lot of ATP/energy from respiration.
- Blood plasma in capillaries around the proximal tubule has a very low water potential compared with the filtrate (due to dissolved blood proteins and reabsorbed ions and glucose).
- This causes water to be reabsorbed into the blood by **osmosis**.
- No urea is reabsorbed.

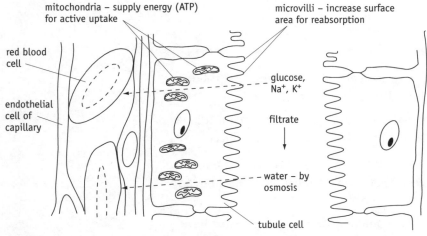

Reabsorption in the kidney tubule

Ⓢ Specific and active reabsorption of substances from the filtrate involves carrier proteins in the membranes of tubule cells. These are specific because of their specific tertiary structures.

Loop of Henle

More ions and water are reabsorbed.

Distal convoluted tubule

More ions and water are reabsorbed, producing **urine**, a concentrated solution of urea.

Collecting duct

This collects urine from many tubules.

- More water can be removed from the urine in the collecting duct.
- The amount of water lost in the urine is controlled by the amount of water reabsorbed by the distal tubule and the collecting duct (see next section).

❶ Make sure you are answering the question set and don't write everything you know about the kidney! If the question asks about reabsorption of glucose or ions, don't write about water!

✓ Quick check 3, 4

❓ Quick check questions

1 Explain why, where and how urea is formed.
2 Describe and explain the differences between blood plasma and the filtrate entering the kidney tubule.
3 Explain how urine is formed from the filtrate entering the kidney tubule.
4 Patients' urine is often tested for the presence of either glucose or protein. Suggest one reason for the presence of either of these in a urine sample.

Regulation of blood water potential

The water potential of blood plasma is kept constant, which keeps the water potential of tissue fluid constant and the same as that of the cell contents – leading to no net gain or loss of water by osmosis. An important factor in regulation of blood water potential is the amount of water reabsorbed by the distal convoluted tubule and the collecting duct.

Reabsorption of water

- If the body has **too much water**, less water is reabsorbed by the distal tubule and collecting duct. A **greater volume** of **dilute** urine is produced.

- If the body has **too little water**, more water is reabsorbed by the distal tubule and collecting duct. A **smaller volume** of **concentrated** urine is produced.

Removing water from the filtrate/urine in the distal tubule and collecting duct is difficult, because it has a very negative water potential. A lot of water has already been reabsorbed by the proximal tubule and loop of Henle.

- To reabsorb more water by osmosis, the surrounding tissue (fluid) has to have a lower water potential than the filtrate/urine.

- The cells of the **loop of Henle** carry out **active transport** of **chloride ions from the filtrate into the surrounding tissue** (fluid) – **which also surrounds the distal tubule and collecting duct**.

- This active transport is against the concentration gradient – maintaining a **gradient of ions** between the filtrate and the surrounding tissue.

- Sodium ions 'follow' the chloride ions. **In effect, sodium chloride is pumped into the surrounding tissue** (fluid), **giving it a more negative water potential than the filtrate/urine**.

- Water can thus be reabsorbed by osmosis from the filtrate into the surrounding tissue and then into the blood.

- How much water is reabsorbed is controlled by changes in the permeability to water of the cells lining the distal tubule and collecting duct.

The longer the loop of Henle, the more 'pumps' there are and the more negative the water potential of the surrounding tissue. Desert-living animals usually have longer loops of Henle, to reabsorb and save more water.

Ⓢ Make sure you understand water potentials. Pure water has a water potential of 0. The more substances there are dissolved in it, the lower the water potential. There is a net flow of water by osmosis from a higher to a lower water potential.

❶ Don't try to learn lots of complex detail about how the loop of Henle works (e.g. counter-current mechanisms) – you won't be asked about it! Just try to grasp the essentials given here.

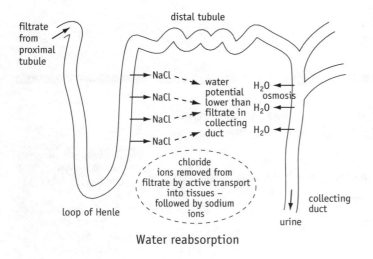

Water reabsorption

Control of the amount of water reabsorption

If the body is losing too much water, the blood water potential becomes slightly more negative.

- This stimulus is detected by receptors in the blood vessels of the hypothalamus, in the brain.
- The response is the release of the hormone **ADH** from the pituitary gland, into the blood.
- ADH makes the cells of the collecting duct and distal tubule more permeable to water; more water is reabsorbed, giving a more concentrated urine.
- If the body has too much water, less/no ADH is released and less water reabsorbed.

? Quick check questions

1 Explain how water is reabsorbed from the urine in the collecting duct.
2 During prolonged exercise a person sweats a great deal and does not drink. Suggest how the body would react to this loss of water.

The eye

The eye is a **sense organ**. It has **receptor cells** called **rods** and **cones** in the **retina**. Too much light damages the retina. The iris controls the light reaching the retina.

- The iris has circular and radial muscles.
- If the circular muscles contract, the pupil gets smaller – less light enters the eye.
- When the radial muscles contract, the pupil gets larger.
- The size of the pupil is changed by a reflex action.

Focusing

A clear image of an object is focused on the fovea of the retina.

- **Most focusing** (refraction, i.e. bending of light) is done by the curved surface of the **cornea**.
- The shape of the **lens** can be altered to carry out **fine focusing – accommodation**.
- When looking at distant objects, the lens is kept thin by **tension** in **suspensory ligaments**.
- To look at near objects, the circular **ciliary muscles** contract, taking tension out of the ligaments – the natural elasticity of the lens makes it fatter (more biconvex).

S Reflexes involve receptors, nerve impulses, sensory neurones, a coordinator, motor neurones, effectors (muscles/glands) and a response.

Remember! The cornea carries out most of the focusing (refraction) of light onto the retina and the lens carries out fine focusing.

Accommodation

✓ *Quick check 1*

Rods and cones

Cones are used for **detailed colour** vision. They:

- are found in the fovea of the retina;
- have a relatively low sensitivity to light – only work in high light intensities.

Rods give **black and white** vision without a great deal of detail. They:

- are found outside the fovea;
- have high sensitivity to light and work in low light intensities.

The structures of rods and cones are different and they contain different light-sensitive pigments.

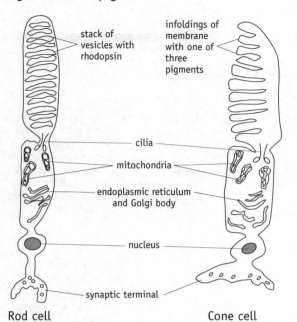

stack of vesicles with rhodopsin

infoldings of membrane with one of three pigments

cilia

mitochondria

endoplasmic reticulum and Golgi body

nucleus

synaptic terminal

Rod cell

Cone cell

ⓢ You need to know what the organelles found in a eukaryotic cell look like when seen with an electron microscope – rods and cones are no exception.

- Rods contain the light-sensitive pigment **rhodopsin** as part of the structure of membranes of vesicles in the outer segment of the rod cell.
- Light 'bleaches' rhodopsin, making it break down to retinene and scotopsin.
- This alters the permeability of the membrane to Na⁺ ions, leading to formation of nerve impulses that pass along the optic nerve to visual centres in the brain.
- In the absence of light, rhodopsin is regenerated from retinene and scotopsin.

◗ Don't learn a lot of unnecessary biochemical detail about the production and turnover of rhodopsin!

✓ *Quick check 2, 3*

? Quick check questions

1 Explain what happens to maintain a clear image on your retina when you change from looking at a distant object to reading a book.

2 At night, we often see moving things out of the corner of our eye, but cannot see them at all when we try to look directly at them. Suggest why this happens.

3 Explain how the stimulus of light falling on a rod cell is converted into information the body can use.

Colour vision

The **trichromatic theory** of colour vision is based upon **three types of cone cell** found in the retina.

- They are blue-absorbing, green-absorbing, or red-absorbing, depending on the wavelength at which they show maximum absorption. Each type contains a **different pigment**.

- Other colours cause reactions in combinations of cones. So, for example, brown light stimulates green-absorbing and red-absorbing cones.

- Red-green colour blindness is caused by a lack of either green-absorbing or red-absorbing cones.

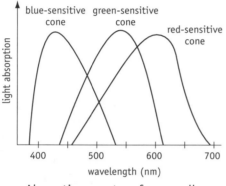

Absorption spectra of cone cells

> **▶** Make sure you write about cone cells with a pigment showing maximum absorption in the red part of the spectrum – **not** 'red cones'!

> **Ⓢ** Colour blindness is due to an allele of a gene which produces a different pigment, compared to the normal pigment. This allele is the result of a mutation.

> ✓ *Quick check 1*

Sensitivity and visual acuity

When we look directly at something, its image falls on the fovea and we see it in colour and sharp detail. Things in the periphery (edge) of our field of view are not seen in colour, or in any detail. The fovea has a very high density of colour-sensitive cones.

- Each **cone** cell has a synapse with one bipolar cell and one ganglion cell.

- Ganglion cells are neurones of the optic nerve, so each cone sends nerve impulses to the brain about its own small area of the retina – giving **high visual acuity**.

Rod cells provide peripheral black and white vision.

- **Rod** cells are connected in groups to one bipolar cell and one ganglion cell.

- Groups of rods send information to the brain about a relatively large area of the retina (compared with individual cone cells).

- If enough light from an object falls anywhere inside the area of a group of rod cells, the brain is aware of this but the view of the object will not be very sharp – the **visual acuity is low**.

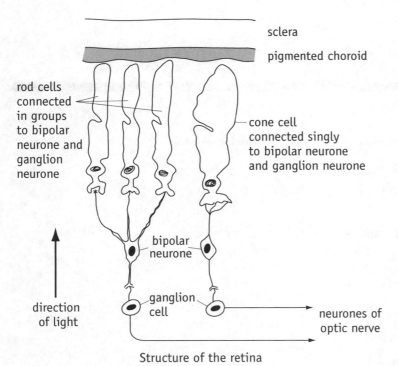

sclera

pigmented choroid

rod cells connected in groups to bipolar neurone and ganglion neurone

cone cell connected singly to bipolar neurone and ganglion neurone

bipolar neurone

direction of light

ganglion cell

neurones of optic nerve

Structure of the retina

> Make sure you understand the links between the ways rods and cones are connected to ganglion cells and visual acuity.

✓ Quick check 2

❓ Quick check questions

1 Some people who are red-green colour blind have three types of cone cell but the pigments of two of the types have very similar absorption spectra. Sketch a graph similar to the one on page 22, to show the absorption spectra of the cones of such a person.

2 Explain why we see so much detail when we look at coloured illustrations.

Nerve impulse

Neurones

Neurones generate **nerve impulses**, carrying information along axons. Neurones communicate at **synapses**, where information is carried by chemical transmitter substances.

Very thin extensions of the neurone carry nerve impulses towards and away from the cell body, which contains most of the cytoplasm and organelles.

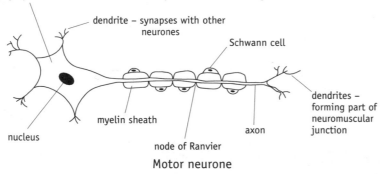

cytoplasm – with mitochondria, endoplasmic reticulum, Golgi bodies, ribosomes

dendrite – synapses with other neurones

Schwann cell

nucleus

myelin sheath

node of Ranvier

axon

dendrites – forming part of neuromuscular junction

Motor neurone

- Motor neurones carry nerve impulses to muscles fibres at neuromuscular junctions, leading to contraction.
- The axon of the motor neurone in mammals is myelinated.
- The myelin sheath of fatty (cell membrane) material is made by Schwann cells; each making a few millimetres of the sheath.

✓ **Quick check 1**

Resting potential

There is a potential difference across the cell membrane of a resting neurone of about **–70 mV** (millivolts) – the membrane is **polarised**.

- This is a due to a higher concentration of **sodium ions** outside the cell.
- The cell membrane is usually relatively impermeable to sodium ions.
- Some sodium ions do slowly diffuse in, but are **actively transported out** by a transmembrane **carrier protein** called a **cation pump (sodium pump)**.
- **Potassium** ions are actively transported into the cell.
- Each type of pump is a protein with a tertiary structure which allows it to recognise and bind to a specific cation.

Ⓢ You have to know the various ways in which substances enter and leave cells across cell membranes. You also need to know how the structure of the membrane is related to its functions.

Action potential

An action potential is a **travelling depolarisation** of the cell membrane; the resting potential disappears for a few milliseconds. An action potential starts at the cell body and travels to a synapse.

- A stimulus causes transmembrane protein channels to open.
- Sodium ions diffuse in along their concentration gradient, causing a rapid, local depolarisation of the membrane.
- The inrush of sodium ions makes potassium channels open, allowing potassium ions to diffuse out and start to restore the resting potential.
- Sodium channels then close and the sodium pump starts to pump sodium ions out of the cell – to restore the resting potential.
- The local depolarisation opens sodium channels in the next section of the membrane – so the action potential travels.

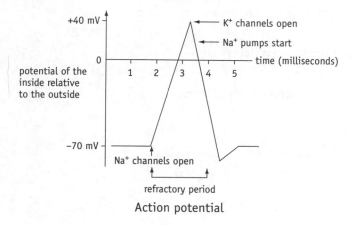

Action potential

- An **action potential** only happens if a stimulus reaches a **threshold** value.
- It is an **'all or nothing event'** – once it starts, it travels to a synapse.
- It is always the **same size** – one travelling action potential is a **nerve impulse**.
- The **frequency of nerve impulses carries information**; a strong stimulus produces a high frequency of nerve impulses.
- After depolarisation, there is a time when no new action potential can start – the **refractory period**. This produces **discrete/separate nerve impulses**.

Invertebrates have **non-myelinated axons**, where nerve impulses travel along the length of the axon. Humans (and other vertebrates) have **myelinated axons**. The axon membrane is exposed at the nodes of Ranvier. Nerve impulses 'jump' from one node to the next, making the speed of travel of the nerve impulses faster than in non-myelinated axons.

> Use the **correct terminology** – many candidates do not use words like depolarised, active transport, sodium channels, tertiary structure, refractory period, resting potential and nerve impulse. Other, vaguer words will **not** get credit.

> The sequence of events involving sodium and potassium ions is important – don't get the two ions confused!

✓ *Quick check 2, 3*

✓ *Quick check 4*

? *Quick check questions*

1 Explain how the structure of a motor neurone is adapted to its function.
2 Explain how an action potential travels along an axon.
3 Use the information in the figure on this page to calculate the maximum number of nerve impulses per second that the axon could carry.
4 We do not react to very small stimuli and there is a limit to our ability to detect the size of very large stimuli. Suggest how these characteristics are linked to the formation and passage of action potentials.

Synapses and drugs

Synapses

Synapses are where neurones communicate with each other, or with an effector, such as a muscle or gland.

- In a synapse there is a gap between the cell surface membranes of the neurones (of about 20 nm) – the **synaptic cleft**.
- Information is carried across the synaptic cleft by **chemical transmitter substances/neurotransmitters**.
- Many different transmitter substances exist; a common example is **acetylcholine** – synapses using this are **cholinergic synapses**.

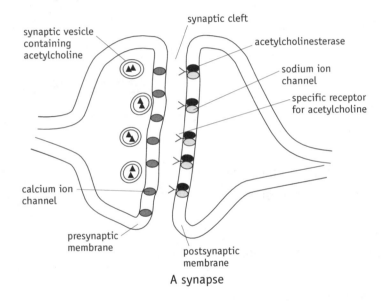

A synapse

- An action potential reaches the **presynaptic** membrane of a synapse.
- The depolarisation causes calcium ion channels in the membrane to open; **calcium** ions diffuse in.
- The calcium ions cause **vesicles** containing acetylcholine to move to the membrane and fuse with it, releasing acetylcholine into the synaptic cleft.
- The acetylcholine **diffuses** very quickly across the synaptic cleft and binds reversibly to **specific receptors** on the postsynaptic membrane.
- The receptors are proteins with a **tertiary structure**/3D shape which allows them to bind specifically to acetycholine.
- Binding of acetylcholine causes sodium channels in the postsynaptic membrane to open.
- Sodium ions diffuse in, causing **depolarisation** of the membrane (an excitatory postsynaptic potential).
- If enough depolarisations are produced frequently enough, an action potential occurs in the postsynaptic neurone.

> Many candidates do not use the correct terms when writing about synapses and miss out important events, such as fusion of vesicles with the presynaptic membrane.

- Acetylcholine in the synaptic cleft has to be destroyed, otherwise it would keep attaching to receptors and causing unwanted depolarisations.
- Acetylcholine is rapidly broken down by an enzyme, **acetylcholinesterase**, which is part of the postsynaptic membrane.

Neuromuscular junctions

Axons of motor neurones reach muscle fibres at neuromuscular junctions. The same events take place here as described for synapses, until acetylcholine binds with its receptor in the **motor end plate membrane**.

- Binding of acetylcholine causes depolarisations of the motor end plate membrane.
- If enough depolarisations happen together, an action potential is created in the cell surface membrane of the muscle fibre.
- This action potential travels through the membrane and causes an inrush of **calcium ions** from the endoplasmic (sarcoplasmic) reticulum and cell membrane into the muscle fibre, triggering contraction.

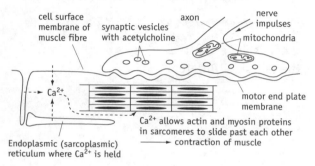

Neuromuscular junction

Effects of drugs on synapses

You are not required to know any specific examples of drugs, so the information needed to answer a question will be in the question.

- Some drugs have a similar shape to a neurotransmitter molecule and compete to bind to receptors on the postsynaptic membrane. This might prevent depolarisation of the membrane, or make depolarisations happen all the time.
- Some drugs bind to acetylcholinesterase and prevent the breakdown of acetylcholine, so that it remains in the synapse, producing continuous depolarisations.

Any of these would spoil the normal communication across the synapse.

? Quick check questions

1 Explain how information is carried across a synapse.
2 Explain why acetylcholinesterase is necessary for the normal functioning of a synapse.
3 Describe the differences between a synapse and a neuromuscular junction.
4 A snake venom paralyses prey. The shape of the venom molecule is shown below. Use the information in the figure on page 26 to suggest how the venom has its effect.

▶ A series of action potentials arriving at a synapse is linked to bursts of release of transmitter substance into the synaptic cleft; these bursts are linked to the number and frequency of action potentials in the postsynaptic neurone.

✓ *Quick check 1, 2*

Ⓢ When looking at synapses, think about the parallels with protein properties linked to tertiary structure and shape fits – enzyme active sites, interaction with substrate, competitive and non-competitive inhibitors.

✓ *Quick check 3*

▶ Look very carefully at any diagrams in questions about drugs and synapses – look for corresponding shapes.

✓ *Quick check 4*

Brain and cerebral hemispheres

Brain

The central nervous system consists of the brain and spinal cord. The brain is made up of:

- the **hindbrain** – part of the **autonomic nervous system**;
- the **midbrain** – processes sensory information;
- the **forebrain** – higher functions, e.g. memory, learning and intellect.

The forebrain

This consists of **left and right cerebral hemispheres**. The left hemisphere controls the right side of the body and the right hemisphere controls the left side. Different parts of the cerebral hemispheres have different functions.

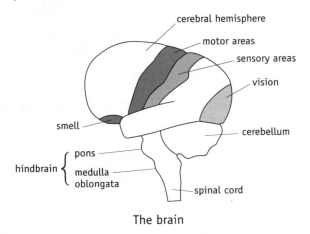

The brain

- **Sensory areas** – receive nerve impulses from receptors. There is a sensory area for each sense, and for different parts of the body in the case of the skin.
- **Association areas** – interpret information from sensory areas and compare it to previous experience/memory.
- **Motor areas** – initiate movement involving voluntary muscles (under conscious control) by sending nerve impulses to muscles (effectors).

The **size** of each sensory, association and motor area depends on the **complexity of its innervation**.

- Humans rely heavily on their sense of sight.
- There are many light receptors in the retina of the eye, producing a large input (complex innervation), requiring a large sensory area (visual cortex).
- This produces complex innervation for a large visual association area.

> You might easily be asked something as straightforward as definitions of these areas. Make sure you know them!

✓ *Quick check 1*

- Humans produce complicated facial movements, involving lots of muscles.
- The motor area devoted to facial movements is very large, with a complex innervation, compared to many parts of the body.

Speech is a special ability of humans.

- There is a large sensory area, the **auditory** area, to receive input from the receptors in the ears.
- This sends information to a large **auditory association area** (Wernicke's area). This connects sounds with appropriate 'things' and words, based on memory and input from other areas of the brain.
- This association area does this job when identifying sounds, or speech, and when we are choosing words to say.
- A large **auditory motor area** (Broca's area) receives information from the association area and coordinates muscle movements to produce speech.
- This involves muscles in the face, lips and tongue.

> ▶ You must use the terms used here when answering a question.

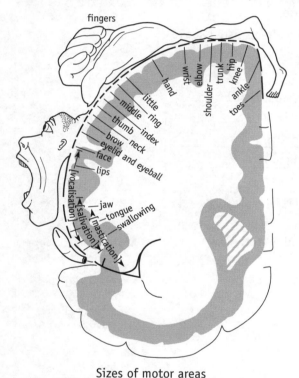

The size of each part of the body as drawn is proportional to the size of its motor area.

Sizes of motor areas

> Ⓢ You might be asked to fit these concepts into an interpretative question concerning stimulus, receptor, coordinator and effector – and the nerve impulses and types of effectors involved.

✔ *Quick check 2, 3*

? Quick check questions

1 Describe the parts played by areas of the brain in the identification of a friend.

2 Use the information in the figure on page 28 to suggest why a doctor tests the eyes of someone who has had a blow to the back of the head.

3 After a stroke, a patient was able to understand and follow spoken instructions but unable to give a spoken reply to a question. Suggest which area of the brain had been damaged by the stroke.

Autonomic nervous system

The autonomic nervous system controls functions that are usually involuntary/non-conscious, including homeostatic systems. Overall control is by the hypothalamus and centres in the medulla (in the hindbrain) of the brain. Local control is through reflexes controlled by ganglia (clusters of neurones) outside the central nervous system.

- The **sympathetic** and **parasympathetic components** of the autonomic nervous system have **opposing effects** on the body.

- The sympathetic prepares the body for 'flight or fight'; e.g. by increasing heart rate and breathing rate, increasing blood supply to muscles, and reducing blood supply to the skin and gut.

- **Noradrenaline** is the neurotransmitter in sympathetic synapses.

- The parasympathetic system uses **acetylcholine** and reverses the effects listed above.

(S) Diverting blood to the muscles gives them more oxygen and glucose for respiration – and more energy for contraction.

Examples of autonomic control   fight or flight.

Target	Symptathetic effect	Parasympathetic effect
Iris of eye	Pupil dilates	Contracts
Bronchi	Dilation	Constriction
Heart	Increase in rate	Slowing of rate
Blood vessels	Vasoconstriction	Vasodilation
Blood supply to gut	Reduced	Increased

Autonomic control involves a stimulus, receptor, coordinator and effector. Two examples are found in the eye.

▶ Make sure that you don't confuse the actions of the sympathetic and parasympathetic systems, or their neurotransmitters.

▶ **Adrenaline** is a **hormone** secreted by the adrenal glands (on top of the kidney). This acts on sympathetic synapses in the same way as noradrenaline – producing very similar effects.

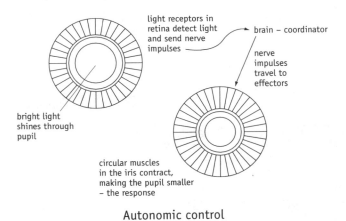

light receptors in retina detect light and send nerve impulses

brain – coordinator

nerve impulses travel to effectors

bright light shines through pupil

circular muscles in the iris contract, making the pupil smaller – the response

Autonomic control

✓ *Quick check 1*

- **Pupil diameter** can be changed to control the amount of light reaching the retina.

- The diameter is changed by circular and radial muscles in the iris.

- If **too much light** (stimulus) reaches the receptors in the retina, the **parasympathetic** nervous system (coordinator) reduces the diameter by making **circular muscles contract** (effectors).

- The sympathetic system increases the diameter of the pupil by making radial muscles contract.

- **Tear production** takes place in response to an adverse stimulus, such as a speck of dust, affecting the cornea.

- The **parasympathetic** nervous system causes the tear (lachrymal) **glands** (effectors) to release tears, to wash out the dust.

Some functions controlled by the autonomic nervous system can be **brought under conscious control**. The **filling and emptying of the bladder** is one example.

- The sympathetic system causes a circular muscle (sphincter) at the opening of the bladder to contract; leading to filling of the bladder.

- As the bladder becomes full, stretch receptors in the walls begin to produce nerve impulses.

- The nerve impulses cause the parasympathetic system to send nerve impulses to the circular muscle, inhibiting its contraction.

- When the muscle stops contracting, the bladder empties (and the stretch receptors stop producing nerve impulses).

- Babies wet their nappies as a result of this system of control. They have to learn conscious control of the circular muscle during 'potty training'.

(S) You might be asked to fit these concepts into an interpretative question concerning stimulus, receptor, coordinator and effector – and the nerve impulses and types of involved.

✓ *Quick check 2*

▶ Examiners will assume that you know these examples!

❓ Quick check questions

1 Suggest how **two** of the examples of sympathetic effects in the table help in a 'fight or flight' reaction to a frightening stimulus.

2 Explain what happens to prevent damage to the retina when we look towards a bright light.

Muscle structure and function

There are three main types of muscle tissue: cardiac, smooth and skeletal. **Cardiac muscle** is in the heart; it does not tire and contractions originate from within the muscle itself. **Smooth muscle** is found in blood vessels and the gut; it contracts slowly and tires slowly. **Skeletal muscle** is attached to bone; it contracts quickly and tires quickly.

Antagonistic muscle action

Muscles concerned with movement are arranged in pairs working in opposition to each other – **antagonistic muscles.**

- When a muscle contracts it shortens and produces a 'pulling' force.
- This contracted muscle returns to its original length by relaxing and being pulled by contraction of another muscle.
- This antagonistic action is used to produce movement, e.g. of arms.

> ◗ Muscles cannot push. Even if you push against something, this action is produced by muscles contracting and pulling!

> Ⓢ At rest these muscles use fatty acids as the main substrate for respiration. They switch to glucose during exercise.

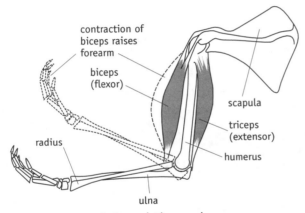

contraction of biceps raises forearm

biceps (flexor)

scapula

triceps (extensor)

radius

humerus

ulna

Antagonistic muscles

- The biceps and triceps are a pair of antagonistic muscles.
- They are attached to the bones by inelastic tendons.
- The biceps (flexor) muscle contracts, raising the lower arm (ulna and radius) upwards.
- At the same time, the triceps muscle relaxes and is lengthened.
- The triceps muscle (extensor) contracts to return the lower arm to its original position.
- At the same time, the biceps muscle relaxes and is lengthened.

✔ *Quick check 1*

Examples of antagonistic muscle action

Most antagonistic muscle action in humans involves movement of bones by the contraction of skeletal muscles. Antagonistic muscle action also occurs in smooth muscle that does not move the skeleton.

- Circular and longitudinal muscles in the gut wall cause peristalsis.
- Circular and radial muscles in the iris of the eye constrict or dilate the pupil.

Structure of skeletal muscle

Skeletal muscle is also known as **striped** or **striated muscle**. Under a light microscope, muscle consists of many long **muscle fibres** or cells.

Muscle fibres:

- are cylindrical in shape and enclosed by a cell surface membrane or **sarcolemma**;
- have many nuclei (**multinucleate**);
- contain numerous protein strands or **myofibrils** with characteristic cross-striations;
- are arranged in parallel, giving a striped appearance as cross-striations line up;
- are surrounded by collagen and connective tissue which extends to form the **tendon** connecting muscle to bone.

Ultrastructure of skeletal muscle

The ultrastructure of muscle can be seen using electron microscopy. The banding pattern of skeletal muscle is due to the arrangement of filaments made from **myosin** or **actin** proteins. Each repeating pattern of banding is called a **sarcomere**.

- Thin filaments are made of actin; thick filaments are made of myosin.
- Thick filaments are present in the **A-band** or **dark band**.
- The outer regions of the A-band are darker, because they contain overlapping myosin and actin filaments.
- The **H zone** at the centre of the A-band is not as dark, because it contains only myosin filaments.
- The M line connects the myosin filaments in the A-band.
- **I-band** or **light band** contains only thin actin filaments.
- The **Z line** connects these actin filaments.

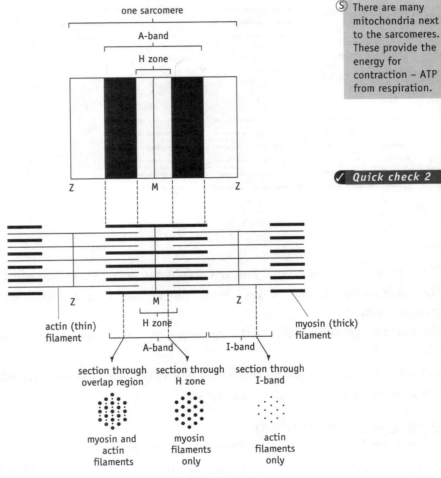

Structure of skeletal muscle

Ⓢ There are many mitochondria next to the sarcomeres. These provide the energy for contraction – ATP from respiration.

✓ *Quick check 2*

✓ *Quick check 3*

? *Quick check questions*

1 Use the information in the figure on page 32 to explain how the left hand can be moved to touch the left shoulder.

2 Name the two types of protein filaments found in muscle.

3 Which part of a sarcomere contains only myosin filaments?

Muscle contraction

The **sliding filament hypothesis** describes the mechanism of muscle contraction. During contraction, thin actin filaments are pulled past and between the thick myosin filaments. This causes shortening of the muscle fibre and changes in the banding pattern – the filaments themselves do not contract or shorten.

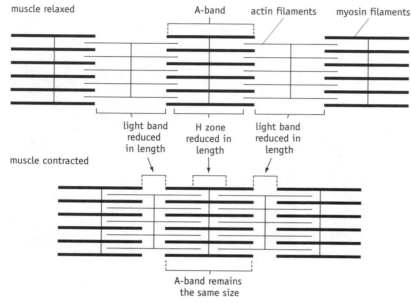

Sliding filament hypothesis

- The **H zone** inside the A-band **narrows** – contains only myosin filaments.
- The **outer darker regions** of the A-band **widen** – contain overlapping actin and myosin filaments.
- The **I-band** (light band) **shortens** – the non-overlapping portion of the actin filaments.
- The **A-band** stays the **same** – the length of myosin filaments does not alter.
- The **Z lines** are pulled towards each other.

> Actin and myosin filaments do not contract, they slide past each other – pulling Z lines closer together.

✓ *Quick check 1*

The ratchet mechanism

This explains muscle contraction and involves the formation of cross bridges between actin and myosin filaments.

- Each myosin filament has a long rod-like region and a myosin 'head'.
- Actin filaments have attachment sites for the myosin heads.
- During contraction, actomyosin bridges form as myosin heads attach to actin filaments.
- Bridges rapidly break and reform along the actin filaments, pulling them past the myosin filaments.
- ATP provides energy for the release of myosin heads from actin.
- ATP is hydrolysed by an ATPase enzyme on the myosin heads.
- Numerous mitochondria supply ATP via aerobic respiration.

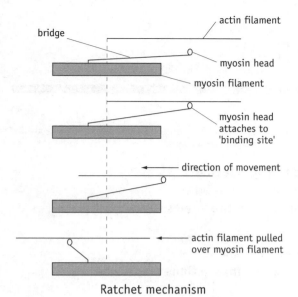

Ratchet mechanism

✓ Quick check 2

Role of calcium ions, troponin and tropomyosin

Muscle contraction is activated by calcium ions released from the sarcoplasmic (endoplasmic) reticulum in the muscle fibre when the fibre is stimulated.

- The binding site on actin filaments in relaxed muscle is covered by the protein, tropomyosin.

- Tropomyosin switches the contraction mechanism on or off.

- Tropomyosin is attached to another protein, troponin.

- When released, calcium ions bind to troponin, causing it and the attached tropomyosin to move from the binding site.

- The actin filament is now switched 'on' and myosin binds to form cross bridges.

- Bridges rapidly break and reform, causing shortening of each sarcomere.

- When the muscle is no longer stimulated, the calcium ions are actively transported back into the sarcoplasmic reticulum.

- Calcium ions stimulate the action of ATPase which hydrolyses ATP, providing the energy for the formation and breakdown of the bridges.

Role of calcium, troponin and tropomyosin

Ⓢ Calcium ions are released when electrical impulses travel from a neuromuscular junction. Nerve impulses travel down the axon of a motor neurone to get to the neuromuscular junction.

Ⓢ The binding of calcium causes a change in the shape of the troponin protein – which changes its function.

✓ Quick check 3

? Quick check questions

1 During muscle contraction, what happens to the length of: (i) the A-band; (ii) the H zone?

2 Explain how energy is provided for the ratchet mechanism during muscle contraction.

3 Explain how calcium ions activate muscle contraction.

Genotype and meiosis

Genotype

A gene is a specific length of DNA carrying coded information for producing a particular protein/polypeptide. It is found at a specific place/**locus** on the DNA *on homologous chromosomes* molecule in a chromosome. **Genotype is the genes an organism has and the alleles of those genes.**

physical

Phenotype is the characteristics of an organism, including the enzymes and substances it produces. It depends on the expression of the genotype – **interactions between the genotype and environment.**

- **Diploid** organisms inherit **two copies of each gene and chromosome;** one from each parent (maternal and paternal).

- **Haploid** cells or organisms have **one copy of each gene and chromosome.**

- A gene can exist in different forms, called **alleles.**

- **Homozygous** – having two identical alleles of a gene.

- **Heterozygous** – having two different alleles of a gene.

- **Dominant** alleles – expressed in the phenotype when homozygous or heterozygous.

- **Recessive** alleles – only expressed in the phenotype in the homozygous state.

Blood groups are codominant.

- Some alleles are **co-dominant** – both are expressed in the phenotype.

A dominant allele usually produces a functional protein and a recessive allele does not. Co-dominant alleles both produce functional proteins.

> The ABO blood groups are determined by 3 alleles of one gene – I^A, I^B and I^0. I^A and I^B are co-dominant and both are dominant over I^0 (a recessive).

look at this example – both parents carry I^0

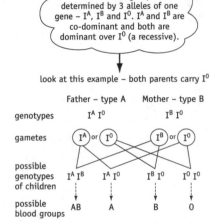

	Father – type A	Mother – type B
genotypes	$I^A I^0$	$I^B I^0$
gametes	I^A or I^0	I^B or I^0

possible genotypes of children	$I^A I^B$	$I^A I^0$	$I^B I^0$	$I^0 I^0$
possible blood groups	AB	A	B	O

Inheritance of ABO blood types: multiple alleles – dominant, recessive and co-dominant alleles

▶ Dominant and recessive have nothing to do with 'better' or 'worse' – or more or less common. Huntington's chorea is a rare, fatal condition caused by a dominant allele. Type O is the commonest blood type in the UK and is homozygous recessive.

Ⓢ A gene is a sequence of bases on DNA, carrying information that can be transcribed into mRNA and translated into polypeptides at ribosomes. New alleles arise by mutation.

✓ *Quick check 1, 2*

Meiosis and fertilisation

Sexual reproduction involves the production of **haploid** male and female **gametes**, which fuse at **fertilisation**. If gametes carried the diploid number of chromosomes or genes, the chromosome number would double at fertilisation. Meiosis **at some point** in the life of an organism keeps the chromosome (and gene) number constant from one generation to the next.

- **Meiosis** produces **haploid cells** with **one copy of each gene.**

- At the start of meiosis, the DNA molecule in each chromosome replicates, forming two identical sister **chromatids**, held together at the **centromere**.

- Chromosomes in diploid organisms form **homologous** pairs, one of maternal and one of paternal origin, carrying the same genes but not always the same alleles.

Ⓢ In some organisms the adults are haploid and produce gametes by mitosis – meiosis occurs after fertilisation.

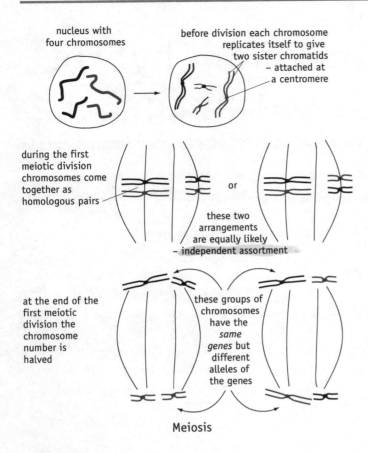

Meiosis

- In meiosis, homologous chromosomes come together and their chromatids wind round each other, forming **chiasmata**.
- Pairs of homologous chromosomes, held together at ~~chiasmata~~ Centromere, are **bivalents**.
- Pieces of chromatid are sometimes exchanged between homologous chromosomes at a chiasma.
- This is **crossing over** which produces **new combinations of alleles.**
- Bivalents line up on the equator of the meiotic spindle and one of each pair moves to each pole of the spindle, **halving the chromosome number**.
- The cell divides, producing **haploid** cells.
- Bivalents can line up on the spindle to give any combination of maternal and paternal origin chromosomes after division – this is **independent assortment**.
- This produces haploid sets of chromosomes containing **new combinations of maternal and paternal origin chromosomes and alleles** and genes.
- The haploid cells' chromosomes still consist of two chromatids, which are separated during a second cell division, to give four haploid cells.

> ▶ Do not try to learn the names and details of all the stages of the first meiotic division – it is not required and no marks are awarded for knowing them.

✓ *Quick check 3*

? *Quick check questions*

1. Explain what is meant by each of the following: (i) genotype; (ii) phenotype.
2. Use the information in the figure on page 36 to explain how parents with type A and type B blood can produce gametes carrying the O allele.
3. Explain why meiosis is necessary at some stage in the life of an organism that reproduces sexually.

Sex determination, monohybrid inheritance and sex-linkage

Sex determination

Sex of humans is determined by sex chromosomes called **X** and **Y**.

- **Women are XX and men XY.**

- Women produce **egg** cells by meiosis; **all contain one X chromosome.**

- Men produce **sperm** cells of **two types**: half contain **one X chromosome**, half contain **one Y chromosome**.

- If a sperm carrying an X fertilises an egg, a **girl (XX)** is produced.

- A sperm carrying a Y produces a **boy (XY)**.

- The chance of having a baby girl is 50% (and 50% for a boy); the father determines the sex of the child.

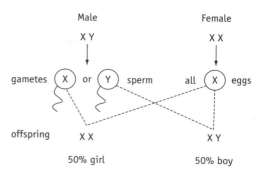

Sex determination in humans

> An amazing number of candidates get confused about XX and XY – XY is male! This mistake can be very costly if the question is about sex-linkage.

✓ **Quick check 1**

Monohybrid inheritance

Monohybrid crosses involve **alleles of a single gene.**

- **Recessive** alleles can cause genetic disorders in humans, e.g. **cystic fibrosis.**

- **Carriers** of the condition are **heterozygous** for the allele and unaffected.

- If two carriers have children, each child born has a 25%, or 1 in 4 chance of inheriting two copies of the cystic fibrosis allele and being affected.

- Questions might involve two generations of crosses, the F_1 and F_2. The example shown on page 39 involves tall and short pea plants.

- Homozygous (pure-breeding) tall plants were crossed with homozygous (pure-breeding) short plants and the F_1 offspring were grown.

- All the F_1 were heterozygous and tall, showing that tall is dominant.

- The F_1 were crossed with each other, producing the F_2 generation.

- The F_2 contained tall and short plants in a **3:1 ratio** – a **phenotypic ratio.**

- The **genotypes** of the F_2 contained 1 homozygous tall : 2 heterozygous tall : 1 homozygous short.

- **Dominant** alleles can cause genetic disorders, e.g. **Huntington's chorea.**

- Most sufferers have a heterozygous parent (sufferer) and a homozygous recessive parent (unaffected).

> A 3:1 ratio is the same as 1 in 4, or 75% to 25%.

> In genetics questions, set out details clearly in diagrams. It is important to read information given in questions; it may help you to find the genotypes of parents in crosses.

> Use a capital letter for a dominant allele, e.g. T, and the small case of **the same letter** for the recessive, t. Do **not** use a different letter!

- 50% of children inherit the disorder but symptoms do not appear until middle age.
- **Co-dominant** alleles are both expressed in the phenotype of heterozygotes; e.g. **ABO blood groups** in humans (see previous section on Genotype and meiosis).

Sex-linked characteristics

Genes on the X chromosome are sex-linked – the Y chromosome has no functional genes on it (as far as you are concerned!).

- Women can be homozygous or heterozygous for sex-linked genes.
- Men inherit **one** copy of a sex-linked gene, on the X chromosome from their mother, which is **always** expressed in their phenotype.
- **Haemophilia** is a recessive sex-linked disorder in humans.
- A boy who inherits one copy of this allele from his mother will suffer from haemophilia.
- This is why haemophilia is much more common in men than women.

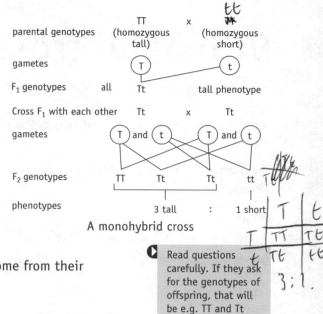

A monohybrid cross

> ▶ Read questions carefully. If they ask for the genotypes of offspring, that will be e.g. TT and Tt and tt. If they ask for phenotypes, the answer will be tall or short.

✓ *Quick check 2*

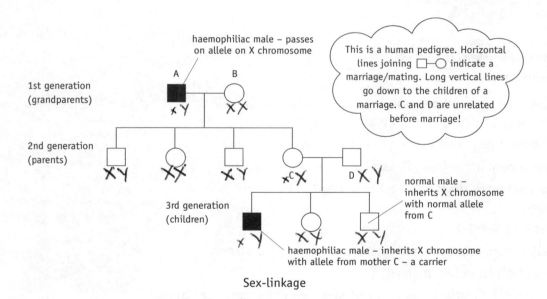

Sex-linkage

> ▶ It is never good enough to say that a gene is sex-linked because more of one sex show it than another. Questions usually give **evidence** that a gene is on the X chromosome – you have to find it! Look at where the X chromosomes are inherited from.

✓ *Quick check 3*

? ## Quick check questions

1 A couple have three sons. Explain their chances of having a daughter if they have another child.

2 A pure-breeding round pea plant was crossed with a pure-breeding wrinkled pea plant. All of the F₁ had round peas. Explain the offspring you would expect in the F₂.

3 A couple whose first child has haemophilia consult a genetic counsellor about their chances of having another affected child. Explain the sort of advice they could be given.

Dihybrid crosses

These crosses involve **alleles of two genes** – the information given in the question is vital.

The example looks at inheritance in pea plants of characteristics of tall or short and yellow or green pea colour.

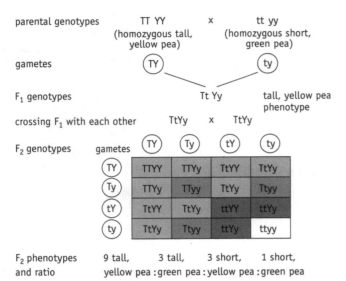

Dihybrid cross

- Homozygous (pure-breeding) tall, yellow pea plants were crossed with homozygous (pure-breeding) short, green pea plants.
- All F$_1$ were tall with yellow peas, showing that tall and yellow are dominant.
- The F$_1$ were crossed with each other (selfed), to produce an F$_2$ generation.
- The F$_2$ contained a ratio of 9 tall, yellow pea : 3 tall, green pea : 3 short, yellow pea : 1 short, green pea.
- This is a typical **F$_2$ ratio of 9:3:3:1**.
- Note how the **Punnett square** is constructed to show the gametes produced.
- Each gamete contains **one** allele of each gene.
- You should be aware that the F$_2$ ratio is the result of **independent assortment** of homologous chromosomes in **meiosis**, leading to gamete production.

> In using Punnett squares, make sure that you indicate which squares represent the gametes (and which parent they are from).

> You may well get an example in a question involving genes which you are not familiar with. The F$_1$ results usually give evidence of which alleles of genes are dominant and which recessive. This evidence should be given in your answer.

> ✓ *Quick check 1*

Epistasis

Epistasis is where one gene affects the expression of another gene. This is seen in **metabolic pathways** controlled by several enzymes, each coded for by a different gene.

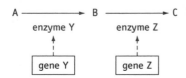

> Ⓢ A is the substrate for enzyme Y which produces B – the substrate for enzyme Z. The rate of action of enzyme Z depends on its substrate concentration.

- To produce C, there must be functional enzymes Y and Z.
- The dominant allele of gene Y produces functional enzyme Y, but recessive allele y does not.
- The dominant allele of gene Z produces functional enzyme Z, but recessive allele z does not.
- Individuals with **genotypes YY or Yy** produce functional enzyme Y and so **produce substance B**.
- If they also have the **genotype ZZ or Zz**, they can then **make substance C from substance B**.
- If they have the genotype zz, then they will only make substance B.
- An individual with the **genotype yy cannot make functional enzyme Y and so cannot make substance B**.
- This means **they cannot make substance C, even if they have ZZ or Zz**.
- In this case, **gene Y has prevented the expression of gene Z – epistasis**.
- The actual example shown is coat colour in mice.

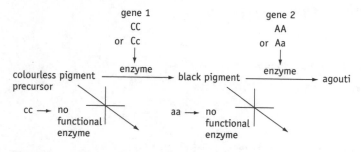

White mice are homozygous cc – it doesn't matter what they are for gene 2
Black mice are CC or Cc for gene 1 and aa for gene 2
Agouti mice are CC or Cc for gene 1 and AA or Aa for gene 2

Epistasis

(S) Non-functional enzymes will have tertiary structures which do not allow substrate to bind to form enzyme–substrate complexes that break down to give products.

- The epistatic gene determines whether the coat is coloured, C, or albino (white), c.
- There is a second gene that determines how any colour is distributed. It exists as agouti (grey), A and black, a.
- The expression of this second gene is affected by the colour gene.
- Epistasis usually reduces the number of phenotypes in the F_2 generation.

✓ **Quick check 2**

? Quick check questions

1 Pea plants homozygous for purple flowers and green pods were crossed with plants homozygous for white flowers and yellow pods. All of the F_1 had purple flowers and green pods. Explain the results you would get if the F_1 were crossed with each other.

2 Melanin is the pigment that makes skin brown. It is produced by a metabolic pathway involving several enzymes. Explain how being homozygous recessive for one gene can produce albinos, people whose skin is always very pale.

Variation

There is a range of types in any population – the individuals are not all identical, even though they belong to the same species. Organisms show a lot of different characteristics and different forms of these characteristics.

Types of variation

Continuous variation

- **Characteristics having a range between two extremes**.
- For example, height and mass of humans show a range from smallest to greatest.
- These characteristics (such as length or mass) can be **measured** in units – they are **quantitative**.
- There are no separate categories or types, only differences of degree.

Discontinuous variation

- **Characteristics showing separate/discrete categories or classes**.
- For example, ABO blood types of humans – only four types and everyone falls into one of the types.
- There are no intermediate types.
- These differences cannot be measured in units – they are **qualitative**.

Results of a monohybrid cross between tall and short pea plants – the height of the F_2 plants was measured

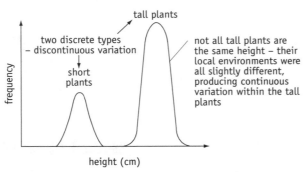

Continuous and discontinuous variation

Causes of variation

There are two main causes of variation between individuals of the same species, **genetic factors** and **environmental factors**. The greatest similarities between individuals occur when they share identical (or very similar) genotypes (identical twins and clones) and are raised in the same environment.

In discontinuous variation:

- there is a **strong genetic factor**;
- often due to **alleles of a single gene**;
- **environmental factors play a small/no part** – e.g. ABO blood groups.

ⓈA population is all the organisms of one species in a habitat. There is intraspecific competition between members of a population. Where there is variation, some phenotypes will be better adapted to the environment and more likely to survive to reproduce.

✓ *Quick check 1*

Sometimes, environmental factors affect discontinuous variation; e.g. tall and short pea plants.

- Genetically tall pea plants grown under different environmental conditions, produce a range of heights but are taller as a group than genetically short plants.

In continuous variation:

- there is a **strong environmental factor(s)**;
- (where genetic factors are involved) **many genes** are involved – **polygenic**.

Many genes contribute to height in humans, but a lack of food, or particular food types, will lead to stunted growth, regardless of the person's genotype.

✓ *Quick check 2*

Identical twins – with same genotypes

	Mean difference	
	Twins raised together – same environment	**Twins raised apart – different environment**
Height (cm)	1.7	1.8*
Mass (kg)	1.9	4.5**
Intelligence (IQ points)	5.9	8.2***

*strong genetic influence
**strong environmental influence
***elements of both genetic inheritance and environment

Sources of genetic variation

When genetic factors are involved in variation, you need to be aware of what produces genetic variation – leading to different genotypes.

- **Mutation** is the source of genetic variation, producing **new alleles** of genes.
- Sometimes they can produce new genes, or combinations of chromosomes.
- **Meiosis** produces genetic variation through:
  - **crossing over** at chiasmata – producing new combinations of maternal and paternal origin alleles on homologous chromosomes;
  - **independent assortment** – producing new combinations of maternal and paternal origin chromosomes and their alleles.
- Both produce genetic variation in gametes involved in sexual reproduction.
- Random fusion of gametes at fertilisation also brings together new combinations of alleles.

▶ Meiosis does not cause mutations! There are some mutations associated with mistakes in meiosis but not ones that concern this course.

✓ *Quick check 3*

? **Quick check questions**

1 Explain whether each of the following is an example of continuous or discontinuous variation: human IQ, haemophilia, length of hair.

2 Identical twins were raised apart and so were non-identical twins. Explain which of the following characteristics you would expect to vary more in the non-identical twins: height, weight, ABO blood group.

3 Explain how meiosis increases genetic variation.

Natural selection

A species exists as populations, groups of organisms of the same species, living in the same habitat and able to interbreed.

Evolution acts on populations, not individuals.

Mutations, meiosis, sexual reproduction and environmental factors produce variation in most populations.

- **Intraspecific competition** happens in a population for means of survival – food, water, mates, space to live or breed.
- **Variation** produces some **phenotypes** that are **better adapted to their environment than others**; they **compete better** – have a **selective advantage**.
- **Differential survival rates** – better adapted phenotypes are more likely to survive, reproduce and **pass on their combination of alleles/genes to the next generation**.
- This **changes the frequencies of alleles and phenotypes** in a population – more advantageous ones increase in frequency, less advantageous ones decrease.
- **Evolution is a change in the frequency of alleles in a population.**
- **Natural selection** is the process by which an **environmental factor(s) affects the survival rates of different phenotypes** in a population.
- **Interspecific competition** is often part of natural selection.
- **Predation** involves competition between predator and prey.
- Some predators are better at catching prey – a selective advantage.
- Some prey are better at avoiding predators – a selective advantage.
- **Disease** involves competition between parasite and host.
- Some host organisms may have a greater resistance to a disease than others.
- Some disease organisms are better at avoiding hosts' defences than others.
- In a **stable environment**, natural selection usually favours the 'average' members of a population – the best adapted to the environment.
- Organisms with extreme forms of characteristics, or mutations, are usually selected against.
- The environment can change; this changes selection pressures on populations.

> Phenotypes are selected for or against, not genotypes.

> Evolution does not have to lead to new species and it has no goal – it is what the definition says it is!

> Ⓢ A population lives in an ecosystem and forms part of a community. Organisms of the population will form part of food chains and webs. Anything that affects the stability of the population will affect other populations of other species.

> ✓ *Quick check 1, 2*

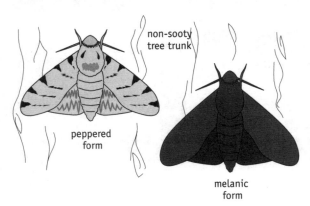

non-sooty tree trunk

peppered form

melanic form

Example – the peppered moth

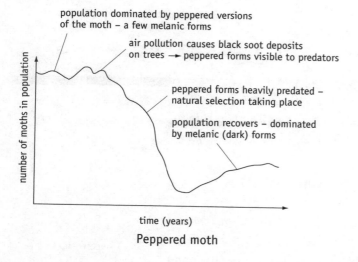

population dominated by peppered versions of the moth – a few melanic forms

air pollution causes black soot deposits on trees → peppered forms visible to predators

peppered forms heavily predated – natural selection taking place

population recovers – dominated by melanic (dark) forms

y-axis: number of moths in population

x-axis: time (years)

Peppered moth

(S) Farming practices often change the environment of populations of other species. Fertilisers, effluent and pesticides all apply new selection pressures to populations of other species.

✓ **Quick check 3**

The commonest mistake that candidates make is thinking that organisms develop a characteristic **because of** a change in the environment; e.g. that introducing a disease forces individuals or populations to evolve. Organisms cannot decide to have a favourable allele or gene – they were either born with it or not!

- For example, pollution changes the environment.

- Some individuals **may have**, **by chance**, a mutation or combination of alleles that gives them better camouflage in the new conditions.

- They have a higher differential survival rate, compared to others in the population.

- This leads to an increase in the frequency of alleles giving camouflage in subsequent generations of the population.

- **Populations do not decide to adapt, or mutate, after an environmental change. The mutation, or combination of alleles giving camouflage, have to already be there by chance, otherwise the population may become extinct.**

? *Quick check questions*

1 Explain why some members of the same population are more likely to survive and reproduce than others.

2 Warfarin is a poison used to kill rats. In some parts of the country the poison is no longer effective, because the local population has become resistant to it. Suggest how these resistant populations developed.

3 In an answer to question 2, a candidate wrote that, 'The rats became resistant to warfarin, so that they could survive'. Explain why this answer is wrong.

Speciation

A **species is a group of organisms that interbreed to produce fertile offspring**. According to the theory of evolution, new species arise from existing species by a process of **speciation**.

Evolution is a change in the frequency of alleles in (or gene pool of) a population. This does not have to produce a new species.

- A species is made up of one or more populations, each living in different (though often very similar) environmental conditions/habitats.
- Natural selection takes place in each population, leading to changes in allele and phenotype frequencies in each population.
- Over time, this process makes each population different from all the others.
- In most cases, individuals move from one population to another – **emigration and immigration** – moving alleles between populations and preventing big differences developing between populations.
- Sometimes a population becomes **isolated**.
- This is usually due to **geographical isolation**; for example, populations living on islands, or different continents, or either side of mountain ranges.
- An isolated population adapts to its environment through natural selection.
- Changes in genotypes and phenotypes in an isolated population **may** lead to **reproductive isolation** and formation of a **new species**.
- The isolated population cannot breed with other populations of the original species to produce viable offspring – they are no longer the same species.

> ◗ You would test to see if two similar organisms belong to the same species by mating them. If they produce viable/fertile offspring, they still belong to the same species.

Darwin's finches

Darwin found different species of finches on different islands of the Galapagos group.

- The species were closely related but obviously different in the food they ate and their beaks.
- Most finches are seed-eaters, with beaks adapted for breaking open the seeds.
- The Galapagos finches have evolved to eat a range of other foods – filling feeding niches usually occupied by other organisms which are not found on the islands.
- For example; on one island finches evolved to catch and eat insects – there were no insect-eating bird species already present.

> Ⓢ In a new environment, a population may find niches that are unoccupied by other species and little interspecific competition for some important factor – such as food. This can lead to adaptive radiation – where the population evolves adaptations that allow it to occupy new niches.

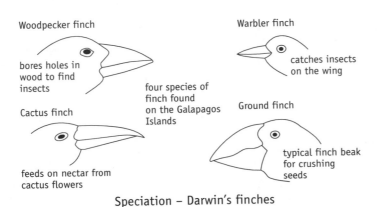

Woodpecker finch — bores holes in wood to find insects

Warbler finch — catches insects on the wing

four species of finch found on the Galapagos Islands

Cactus finch — feeds on nectar from cactus flowers

Ground finch — typical finch beak for crushing seeds

All four species evolved from one ancestral species. A small population of this species colonised the Galapagos Islands.

Speciation – Darwin's finches

Evolutionary change over a long period of time has resulted in the great diversity of forms we see among living organisms. Existing species have arisen from previously existing species by evolution – leading to speciation. Most species that have ever existed are extinct.

- Fossils give evidence of changes in forms of life over hundreds of millions of years and vast numbers of now extinct species.

- They also give evidence of development of more complex forms of life with time.

- Some fossil records show the evolution of new/modern species from previously existing species – for example, the fossil history of the horse.

- Fossils also show structural similarities between extinct and living species.

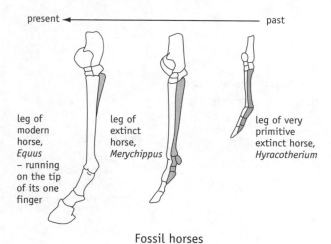

present ◀————————————————————— past

leg of modern horse, *Equus* – running on the tip of its one finger

leg of extinct horse, *Merychippus*

leg of very primitive extinct horse, *Hyracotherium*

Fossil horses

Ⓢ It can be difficult to classify fossils. They may be of organisms with no living relatives – which means a new taxonomic group has to be created. Most fossils also give little information about some important characteristics – like how the organism reproduced, its physiology and its behaviour.

✓ *Quick check 1,2,3*

▶ Complete fossil records like that of the horse are very rare, because very few fossils are formed compared to the numbers of species and individuals that have existed.

❓ *Quick check questions*

1 Explain how the finches on the Galapagos Islands evolved into a new species.

2 a Using the information in the figure on this page, describe the changes that have taken place in the leg bones of horses during evolution.

 b Horses evolved as fast-running animals of grasslands. Suggest how the changes in their leg bones adapted them to their way of life.

3 An exam candidate wrote that, 'The purpose of evolution is to create new species'. Explain why their answer is incorrect.

Classification

The science of classification is known as **taxonomy**. Part of this science is **nomenclature** – the naming of organisms.

- The **binomial system** is used for scientific naming of species.
- Each name consists of **two** words, the **generic** name and the **specific** name.
- The generic name starts with a **capital** letter and the specific name starts with a **small case** letter.
- The name is either **underlined, or written in italics**.
- The words used in the name are forms of Latin/Greek.

In classification, organisms are put into **taxa**. A taxon contains organisms which share some basic feature. The different levels of taxon are: **kingdom, phylum, class, order, family, genus, species**. These taxa make up a **hierarchy**.

- The **species** is the **smallest** taxon; it contains only one type of organism.
- A **genus contains** one or more species.
- A **family contains** one or more genera.
- An **order contains** one or more families.
- A **class contains** one or more orders.
- A **phylum contains** one or more classes.
- A **kingdom** is the **largest** taxon and **contains** one or more phyla.

> One hierarchy is the complete classification of one species – from species to kingdom.

A kingdom is a **composite group**, made up of phyla; a phylum is a composite of classes, and so on until the species is reached – which is not a composite group. The hierarchy means that any organism will have a unique classification. It fits into a certain set of taxa, because there are **no overlaps** between taxa.

Human classification:

kingdom	Animalia
phylum	Chordata
class	Mammalia
order	Primates
family	Homonidae
genus	Homo
species	*Homo sapiens*

Phylogenetic classification

Phylogenetic classification reflects evolutionary links between organisms and their evolutionary history.

- Species in the same genus share a common ancestral species, from which they evolved.
- Genera in a family share a common ancestor, but further back in evolutionary history than the species in a genus.

- Families in an order also share an ancestor, but even further back in evolutionary history.

This type of relationship applies all the way up the hierarchy, to the kingdom.

✓ *Quick check 1, 2*

The five kingdoms

Modern classifications recognise five kingdoms: Kingdom Prokaryotae, Kingdom Protoctista, Kingdom Fungi, Kingdom Plantae and Kingdom Animalia.

Kingdom	Distinguishing features
Kingdom Prokaryotae	Bacteria and blue-green bacteria – microscopic, prokaryotic cells
Kingdom Protoctista	Organisms with eukaryotic cells that are not in the Kingdom Fungi, Plantae or Animilia – often called Protozoa and Algae – unicellular, filamentous, colonial or macroscopic (e.g. seaweeds)
Kingdom Fungi	Eukaryotic with cell walls made of chitin and no cilia or flagella – some unicellular, others consist of thread-like hyphae which form a mycelium – cannot photosynthesise and are saprotrophic or parasitic
Kingdom Plantae	Eukaryotic with cell walls made of cellulose, large vacuoles and chloroplasts (certain cells) – multicellular and photosynthesising – adapted for life on land (most) – growth restricted to meristems (layers/patches of dividing cells)
Kingdom Animalia	Eukaryotic with no cell walls – multicellular – obtain nourishment by feeding, cannot photosynthesise – nervous system – growth throughout tissues (no meristems)

Ⓢ You dealt with the differences in cell organelles in prokaryotic and eukaryotic cells in AS Module 1. The main differences are the much smaller size of prokaryotic cells and their lack of membrane bound organelles.

✓ *Quick check 3*

Viruses

Viruses consist of nucleic acid and a protein coat – they have no cellular structure. They are not thought of as being alive in the usual sense and are not included in most biological classifications.

? *Quick check questions*

1 Explain what is wrong with the following statement: 'The correct name for humans is Homo Sapiens'.

2 Recent discoveries in genetics have shown that humans share 98% of their genetic material with chimpanzees. Suggest how phylogenetic classification can explain this.

3 Describe the differences between a plant and a fungus.

Module 4: end-of-module questions

1 a   i   Give two products of the light-dependent reaction. *ATP NADPH* [2]

      ii   In which part of the chloroplast does the light-dependent reaction take place? *grana* [1]

b   The diagram shows part of the light-independent reaction.

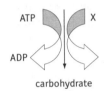

ribulose bisphosphate $\xrightarrow{\text{CO}_2}$ 2 x glycerate 3-phosphate

ATP ⟶ ⟵ X

ADP

carbohydrate

How many carbon atoms are present in a molecule of:

   i   ribulose bisphosphate *5*

   ii   glycerate 3-phosphate? *3* [2]

c   Name molecule X and explain its role in the light-independent reaction. [2]
NADPH used to reduce the glycerate 3-phosphate.

d   How is ribulose bisphosphate regenerated in the light-independent reaction? *Some of the sugar is regenerated by the calvin cycle using ATP.* [1]

2   The diagram shows the main stages of aerobic respiration.

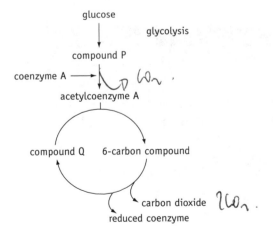

glucose

glycolysis

compound P

coenzyme A ⟶ *CO₂*

acetylcoenzyme A

compound Q   6-carbon compound

carbon dioxide *2CO₂*

reduced coenzyme

a   In which part of the cell do the following processes take place:
(i) glycolysis; (ii) Krebs cycle? [2]
cytoplasm.   ~~mitochondrion~~ *mitochondria*

b   How many carbon atoms are present in a molecule of: (i) compound P; *3*
(ii) compound Q? *4* [2]

c   Explain how reduced coenzyme is used to form ATP. [3]
donates electron to the electron carrier chain to give ATP.

3 a What is meant by homeostasis? *Maintaining the internal body environment at a constant.* [1]

b Describe the role of hormones in the control of blood glucose concentration in a mammal. *Insulin turns blood glucose into glycogen for storage in muscles* [6]

c Describe the process of deamination. [3]

4 The table shows the absorption of light of different wavelengths by the three different types of cone cell.

Wavelength (nm)	Light absorption as percentage of maximum		
	Blue 'sensitive'	Green 'sensitive'	Red 'sensitive'
425	100	0	0
550	0	85	85
600	0	25	75

a Use information from the table to determine which colour would be perceived when light is seen of wavelength: (i) 425 nm; (ii) 600 nm. [2]

b What colour is perceived when there is equal stimulation of all three types of cone cell? [1]

c Explain how rod cells enable some animals to see clearly at low light intensities. [2]

d Describe the changes which occur in the eye when a man looks at his watch after viewing a distant object. [5]

5 a Describe how a resting potential is maintained across the surface cell membrane of a neurone. [3]

b Explain what is meant by the all-or-nothing nature of an action potential. [2]

c Describe the process of synaptic transmission at a neuromuscular junction. [5]

6 Grazing mammals, such as antelopes, rapidly take flight at the sight of a predator, such as a lion.

a Describe how the different areas of the cerebral hemispheres are involved in this response. [3]

b When an antelope sees a predator, its pupils dilate. Describe the role of the nervous system in this response. [4]

7 The diagram shows part of a skeletal muscle.

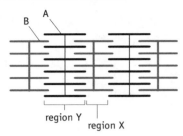

a Name the filaments labelled A and B. [2]

b When a muscle contracts, what happens to the length of:
(i) region X; (ii) region Y? [2]

c Explain the mechanism of muscle contraction. [6]

8 The diagram shows a plant cell with eight chromosomes.

a Draw a diagram to show the chromosomes present in a gamete
produced from this cell by meiosis. [1]

b Give two ways by which meiosis produces variation in gametes. [2]

9 The table shows the phenotypes and possible genotypes of ABO blood groups.

Blood group phenotype	Blood group genotype
A	$I^A I^A$, $I^A I^o$
B	$I^B I^B$, $I^B I^o$
AB	$I^A I^B$
O	$I^o I^o$

a I^A, I^B and I^o are all alleles of the same gene.

i Explain what is meant by an allele. [1]

ii Which allele is recessive? [1]

iii Which alleles are co-dominant? [1]

b Complete the genetic diagram to show all the possible ABO
blood groups of the children of parents of blood group A (heterozygous) and
B (heterozygous).

parental phenotypes   blood group A   blood group B

parental genotypes

genotypes of gametes

genotypes of children

phenotypes of children [3]

10 Madagascar is a large island which broke away from Africa at least 120 million years ago. Following this many new species evolved and the island has been described as the laboratory of evolution.

   a Explain what is meant by the term species. [2]

   b Explain the process involved which might have led to the evolution of new species on the island. [4]

   c Briefly describe the principles on which the classification of the new species is based. [3]

Module 5: Environment

This module is broken down into five topics: Energy flow through ecosystems; Materials are recycled in ecosystems; Studying ecosystems; Dynamics of ecosystems; Human activities and the environment.

Energy flow through ecosystems

- Photosynthesis is the major route of entry of energy into ecosystems.
- Energy passes along the food chains and webs of ecosystems from one trophic level to the next.
- Energy is lost at each trophic level, mainly as heat from respiration.
- The numbers of organisms, amount of biomass and energy in each trophic level can be represented in pyramids of number, biomass and energy.

Materials are recycled in ecosystems

- Important chemical elements are recycled; they are released from living organisms, or their dead remains, and used by other organisms.
- Microorganisms are very important to recycling.
- Decomposers use dead organisms for respiration, releasing carbon dioxide to be used by plants in photosynthesis.
- Bacteria are important in the recycling of nitrogen.

Studying ecosystems

- To study an ecosystem there are techniques for recording and measuring the position and numbers of organisms of the biotic environment.
- Abiotic factors such as pH, light and temperature are measured, because they influence the distribution of organisms.
- Statistical tests discover whether differences between sets of data are significant or not.

Dynamics of ecosystems

- An ecosystem has a community made of populations of different species, living in the same habitat.
- Each species has its own niche and feeding relationships.
- Relationships between species are dynamic – they change with time.
- Populations vary in size due to competition and changes in abiotic factors.
- Succession is when the community in an area changes with time, giving rise to a new community, until a climax community is reached.

Human activities and the environment

- Human activities have a large impact on the environment.
- Farming practices such as use of monocultures and removal of hedgerows reduce biodiversity – the numbers of species and individuals present.
- Organic effluent and nitrate and phosphate based fertilisers damage aquatic ecosystems.
- Pesticides used in agriculture can bioaccumulate in food chains, harming organisms which do not harm crops and may be useful to farmers.
- Some pesticides might get into human food chains.
- Balance is needed between the need to provide food for human populations and conservation of other species, habitats and communities.

Energy flow through ecosystems

Photosynthesis is the means by which energy enters an ecosystem. Plants are **autotrophs**, making biological molecules from sugars produced in photosynthesis and inorganic mineral ions.

- **Primary producers – plants –** provide food for all other organisms.
- **Food chain –** a series of feeding relationships – starts with a primary producer.

tertiary consumer (carnivore)	fourth trophic level
⇑	
secondary consumer (carnivore)	third trophic level
⇑	
primary consumer (herbivore)	second trophic level
⇑	
primary producer (plants)	first trophic level

- **Feeding transfers energy –** arrows point in the direction of energy flow.
- Each level in the food chain is a **trophic level** (energy level).
- Little energy is transferred from one trophic level to the next, because energy is **dissipated (lost)** by organisms at each trophic level through:
  - **respiration,** which provides energy used by organisms for, e.g. movement, growth, reproduction;
  - **heat** energy from respiration;
  - **excretion** of waste products like carbon dioxide and urea;
  - **decomposition** of dead organisms which were not eaten.
- **Decomposers –** bacteria and fungi feed on dead organisms.
- Eventually, all energy is **dissipated back into the environment** as heat energy.
- **Efficiency of energy transfer between trophic levels** can be calculated.

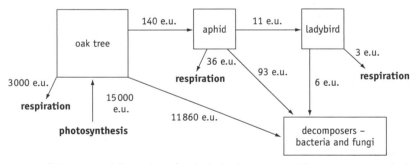

Some of the energy flows in a food chain (e.u. are arbitrary energy units)

Example. What percentage of the energy input from photosynthesis goes to decomposers?

Input from photosynthesis = 15 000 e.u.

Energy input into decomposers = 11 860 + 93 + 6 = 11 959 e.u.

% of energy going into decomposers = $\dfrac{11\ 959}{15\ 000} \times 100 = 79.7\%$

S In photosynthesis light energy is converted to energy in chemical bonds holding sugar molecules together. All organisms carry out respiration all the time to produce ATP. Heat energy is lost during the process.

▶ Some organisms occupy more than one trophic level. Humans are omnivores – they eat plants and animals.

S Fungi secrete digestive enzymes onto their food – extracellular digestion. The soluble products are taken up across their cell membranes.

✓ *Quick check 1, 2*

- Food chains are part of **food webs – interconnected food chains.**

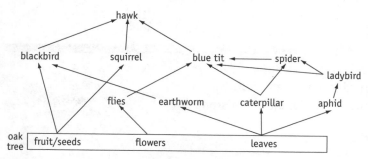

Simplified food web

Ecological pyramids

Pyramid of number – **number of organisms** at each level of a food chain/web **at a specific time.** Primary producers are the base of the pyramid. The size of a block represents the number of individuals at each level in a given area.

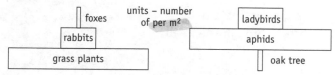

Pyramid of number

- This often gives a misleading impression of the energy transfers. For example, one oak tree will support very large numbers of consumers.

Pyramid of biomass – the **mass of organisms** in each level of a food chain at a **specific time**, **in a given area** (e.g. $kg\,m^{-2}$).

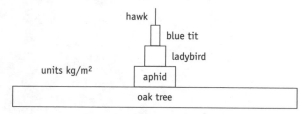

Pyramid of biomass

- This gives some idea of energy present but it does **not** give the **rate** at which matter/energy passes between trophic levels.

Pyramid of energy shows **energy present** at each trophic level, for a **particular area**, for a **particular period of time** (e.g. $kg\,m^{-2}\,y^{-1}$).

✓ *Quick check 3, 4*

? Quick check questions

1 Suggest why energy losses mean that food chains can only involve five or six trophic levels at most. ~~XXXX~~ *Only a small amount of energy is transferred from one trophic level to another. Not enough to sustain another trophic level.*

2 What percentage of the energy input from photosynthesis goes to ladybirds? *0.07%*

3 Explain the differences in units used in pyramids of number, mass and energy.

4 Suggest why it is difficult to get data for pyramids of energy. *not certain specifically for that year.*

Nutrient cycles

In an **ecosystem** interactions occur between the **biotic** (living) and **abiotic** (non-living). Chemical elements constantly enter and leave the biotic environment.

The carbon cycle

- **Carbon** enters the biotic environment when **carbon dioxide** (inorganic) is used to make **sugars** during **photosynthesis**.
- It returns to the abiotic environment as carbon dioxide, during **respiration**.
- **Microorganisms decompose** dead remains of organisms, returning **inorganic ions** to the environment for other organisms to reuse.
- **Bacteria and fungi** are **microorganisms** and important **decomposers**, obtaining their food by **saprotrophic nutrition**.
- They secrete digestive enzymes onto their food to digest large, insoluble food molecules into smaller, soluble molecules that they can absorb through their cell surface membranes.
- Many inorganic ions are in short supply in the abiotic environment and are limiting factors in the growth of producers.
- **Microorganisms** carry out **respiration**, releasing carbon dioxide to the atmosphere.

Ⓢ You need to know how plant roots take up water and mineral ions.

Ⓢ Decomposition happens faster at warmer temperatures, because the enzymes of bacteria and fungi work faster – up to their optimum temperature.

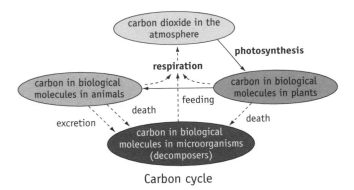

Carbon cycle

✓ *Quick check 1*

The nitrogen cycle

- Plants need the element nitrogen for **amino acid and protein synthesis** (and nucleotide synthesis).
- Plants cannot use molecular nitrogen gas from the atmosphere.
- **Nitrogen fixation** – atmospheric nitrogen is converted into forms plants can use (ammonia, and **nitrate ions (NO_3^-)**), by **nitrogen-fixing bacteria**.
- *Rhizobium* is a bacterium in root nodules of leguminous plants (e.g. clover); it fixes nitrogen into ammonia, which the plant can use.
- **Nitrifying bacteria** convert ammonia to nitrites and nitrates – **nitrification**.

❶ Do not try to learn the proper names of all the various types of bacteria in the nitrogen cycle.

- **Denitrifying bacteria** break down nitrates, releasing nitrogen gas – **denitrification**.

- Consumers in food chains depend on plants for their amino acids.

- Animals convert surplus amino acids to ammonia and organic acids – **deamination**.

- Ammonia (or urea or uric acid (nitrogenous wastes)) is excreted.

- Ammonia is produced when dead organisms are **decomposed**.

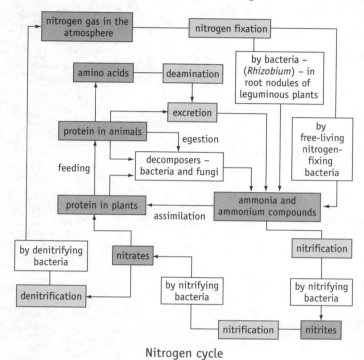

Nitrogen cycle

✓ *Quick check 2, 3*

? *Quick check questions*

1 Explain the significance of photosynthesis and respiration in the carbon cycle.

2 Explain the importance of microorganisms in the recycling of materials in ecosystems.

3 **a** Explain how nitrogen in nitrogen gas in the atmosphere gets into proteins in plants.

   b Suggest why nitrates in the soil are often the limiting factor in the productivity of an ecosystem.

Ecological techniques

Line transects and frame quadrats

- **Line transects** study changes in vegetation types from one area to another.
- A tape measure/string is stretched across areas being studied.
- Plant species are recorded at regular intervals along the tape/string; perhaps using a frame quadrat.
- A **limitation** is that one transect may not cross typical areas, so they are **repeated several times**, along random lines and the results averaged.
- A **frame quadrat** is of known area (usually 0.25 m^2 or 1 m^2).
- **Percentage cover** – by counting the quadrat squares a plant is covered by.
- **Limitations** of the frame quadrat:
  - difficulty in estimating the number of squares covering each species;
  - the small area of the quadrat, compared with the area being studied.
- Frame quadrats can be:
  - placed next to a line transect at regular intervals;
  - used to find percentage cover of a plant in an area, by throwing it at random (to avoid picking 'interesting' areas);
  - placed at points on a grid chosen using random number tables.
- Results are taken from many (20+) quadrats and the results averaged.
- **Abiotic factors** are important, because physiological adaptations of organisms only allow them to live in a certain range of e.g. pH, light and temperature – it is part of what defines their niche.
- pH of soil or water is measured using a pH meter, or universal indicator.
- Light is measured with a light meter, and temperature with a thermometer.
- **Limitations** – factors depend on time of day, weather and season. Readings over a long period of time are needed for representative results.

Statistics

Standard deviation measures spread of data about the mean.

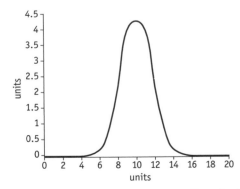

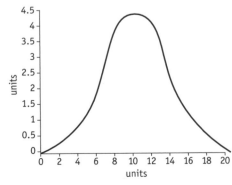

Standard deviation

The ecosystem described in a question might not be one you have studied, but the methods you have used in your studies will still apply.

100 small squares – each equal to 1% of the area of the quadrat

Many candidates give poor answers to 'methods' questions, because they fail to give details – e.g. what exactly is recorded, units, how often, the need for many results/repetitions and averages.

✓ *Quick check 1, 2*

Ⓢ Animals respond to stimuli – including changes in their abiotic environment – with behavioural changes that increase their chances of survival.

Standard deviation is a measure of the variation of a characteristic.

- The sets of data in the graphs on page 60 have the same mean, but the data on the left has a smaller standard deviation.

Chi-squared (χ^2) is used to decide if differences between sets of data are significant.

$$\chi^2 = \Sigma \frac{(O - E)^2}{E}$$

- **O** are **observed results** – what was measured or recorded.
- **E** are **expected results** – calculated from observed results.
- **Null hypothesis** – there are **no significant differences between sets of data**.

Example: Voles (small herbivorous mammals) were trapped at four different sites.

	Trapping site			
	A	**B**	**C**	**D**
Number of voles	33	29	8	30

- **To find the expected results** (based on the null hypothesis), add the observed results together and divide by the number of sets of data.

33 + 29 + 8 + 30 ÷ 4 = 25

Site	O	E	(O–E)	(O–E)^2	$\frac{(O - E)^2}{E}$
A	33	25	8	64	2.56
B	29	25	4	16	0.64
C	8	25	−17	289	11.56
D	30	25	5	25	1.00

$$\chi^2 = \Sigma \frac{(O - E)^2}{E} = 15.76$$

- There are tables of values for χ^2, for different degrees of freedom; in this example there are three degrees of freedom (one less than the number of sets of data).
- A **probability value** of χ^2 **less than 0.05**, means a **significant difference** between observed and expected results.
- The χ^2 value of 15.76, for three degrees of freedom, comes between probability values of 0.01 and 0.001.
- The null hypothesis is rejected – there are significant differences between the numbers of voles trapped at different sites.

✓ *Quick check 3*

? *Quick check questions*

1 Explain how you would investigate the changes in vegetation as you walked from a field into a wood. belt lihe transect. every 5 m record % cover.

2 Suggest how you would investigate the abiotic factors that oak trees can live in.

Ecosystems

An **ecosystem** is a **community** of populations of species and their abiotic environment.

Term	Definition
Population	All the organisms of one species living and interbreeding in a habitat
Habitat	Where an organism lives, including its abiotic and biotic environment
Abiotic factors	Physical conditions; e.g. temperature, pH, rock, soil particles, wind
Biotic factors	The influences due to other organisms
Community	Consists of populations of species in the same habitat – often named after the dominant plant, e.g. oak wood
Niche	Where an organism is found in the habitat and its role in its community – described in terms of the range of abiotic factors it needs, and feeding requirements

You must **learn** the definitions in the table!

- A community has **energy** flowing through its **food web**.
- **Feeding relationships are dynamic** – energy transfer changes with time, with changes in numbers of organisms in each trophic level.
- Species have evolved **adaptations** to survive and reproduce in a niche.
- For example, woodlice are found in dark, damp places – detritivores feeding on dead leaves – this describes their niche.
  - **Structural adaptation** includes a flattened body, to get under stones, logs, etc and mouthparts adapted for cutting up dead leaves.
  - **Physiological adaptations** include enzymes for digesting food.
  - **Behavioural adaptations** to find dark, damp conditions (abiotic factors).

✓ *Quick check 1*

Stability of populations

The populations an ecosystem can support change, due to changes in abiotic and biotic factors. Growth of a population stops due to **limiting factors**.

Abiotic factors affect the size of populations, but are often not greatly changed themselves by sizes of populations (density independent factors).

Climatic factors include seasonal changes in temperature, day length and rainfall, and longer-term changes due to climate changes – natural, or affected by human activity, e.g. global warming.

Inorganic ions such as **nitrate** and **phosphate** often limit plant growth.

Plants are **primary producers**, affecting all populations in a community.

Biotic factors – interactions between organisms

- **Intraspecific competition** occurs between members of the same species.
- As a population grows, (density dependent) competition increases for space (e.g. a patch of soil to grow on or a nesting site) and food.

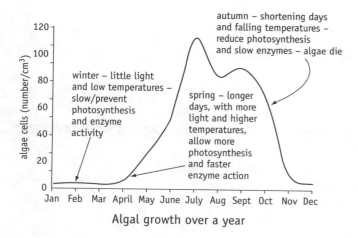

Algal growth over a year

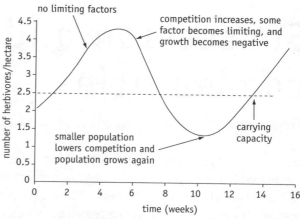

Intraspecific competition in a population of small herbivores

- Population varies about the carrying capacity – an average figure.
- **Interspecific competition** – different species compete for the same resource, at the same trophic level.
- Species of plant compete for light; herbivore species compete for the same plant; or carnivore species compete for the same prey.
- Predation – the predator is a limiting factor on growth of the prey population and the prey is a limiting factor on the predator population.

Ⓢ There will be variation in phenotypes of a population due to genetic and environmental factors. Mutation, meiosis and sexual reproduction produce genetic variation. Variation means that some are better adapted to their environment.

⏵ On a graph like this, always look at the scales on the vertical axes – they are often very different. This may affect any comparisons you are asked to make.

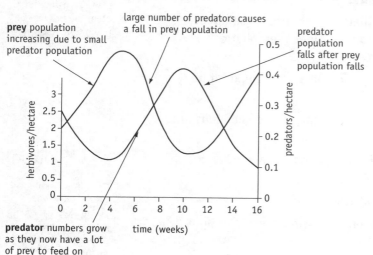

Predator–prey relationships

- Population curves for herbivore and predator have the same general shape.
- The predator's curve always shows a **time lag**, compared to the prey.

✓ Quick check 2, 3

❓ Quick check questions

1 Describe *three* adaptations that humans have to enable them to survive in their environment.

2 Suggest and explain how *one* biotic factor could explain the changes in the size of the population of microscopic algae in the graph on this page between July and September.

3 Suggest what would happen to the populations shown in the graph of predator–prey relationships if an organism was introduced which fed on the same food as the prey organism.

Succession and climax communities

Areas of bare land occur where new soil is formed, or where a community has been destroyed. If bare land is undisturbed, a process of **succession** will take place. Communities will form and be replaced by later communities, until a **climax community** is established which remains stable over a long period of time. A **sand dune succession** is described here, but the same principles would apply to any other succession you have studied.

- The first community on bare land consists of **colonisers**, mainly species of **herbaceous (non-woody) plants** and the organisms that feed on them.

- These species are able to establish themselves quickly and are adapted to withstand unfavourable abiotic environmental factors.

- In sand dune complexes, new dunes are formed as small mounds of loose sand deposited by the wind. These 'embryo dunes' give poor anchorage for plants; the sand does not hold water and is poor in nutrients.

- Marram grass is a species of herbaceous plant that can colonise these dunes, because it has many adaptations to living in dry conditions (xerophytic adaptations).

- Over time, organisms of the 'marram community' die and are decomposed by bacteria and fungi.

- This adds organic material – humus – to the sand, and nitrogen compounds are returned to the soil.

- Humus improves the water-holding capacity of the soil and binds the sand particles together more firmly.

- These changes in the soil make the environment less severe.

- Plants also change abiotic factors such as temperature, wind speed and light by providing shade and acting as wind breaks.

- They also produce new niches for new species to enter and form a new community, such as grasses.

- These grow fast and close together, **out-competing** the colonisers to form a 'grass community'.

- The grass community continues to alter the environment, making it possible for a 'shrub community' to become established – dominated by species such as bramble or sea-buckthorn.

- Finally a 'tree community' replaces the shrub community – dominated by species such as birch or oak.

- Changes from one type of community to the next happen because of inter-specific competition.

- For example, trees finally dominate due to their height and ability to exclude light from shrubs and most herbaceous plants.

The changes and improvements in soil factors are very important and are brought about by plant communities. The availability of water and mineral ions is critical to plant growth, and other organisms rely on plants!

Humus holds water and dissolved mineral ions which can be recycled. This makes water more available for plant roots to take up by osmosis along a water potential gradient. More mineral ions are available for uptake by active transport. Higher temperatures give faster rates of enzyme activity and faster rates of respiration – providing more energy for growth, reproduction, etc.

The climax community that finally forms depends on abiotic factors. **A succession can stop before it reaches a tree community**. This is usually due to abiotic factors which are too great for a community to alter greatly. An example is seen in the changes in plant communities as you go from a valley to the top of a high mountain.

✓ *Quick check 1, 2*

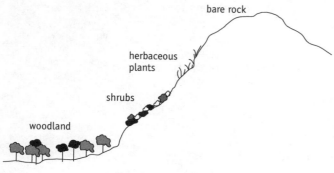

bare rock

herbaceous plants

shrubs

woodland

Climax communities

- In the valley, the abiotic factors are not severe and succession goes through to a climax woodland community (like oak).

- Higher up it is too cold, exposed to wind, or the soil is to poor to allow tree seedlings to grow; succession stops at a climax community dominated by shrubs (like heather).

- Higher still, conditions are too extreme for shrubs, and the climax community is made up of herbaceous plants (like grasses).

- On the mountain top the conditions are too extreme for plants, and bare rock is found.

▶ A question may concern a different succession to the one you studied. You have to apply the ideas you learnt to the new example. Do not ignore the question and write about what you studied!

✓ *Quick check 3*

? *Quick check questions*

1 Describe what is meant by a succession.

2 Explain how colonisation of bare soil by herbaceous plants can make it possible for woody plants to grow.

3 Suggest why marram grass does not grow on old sand dunes which have a woodland community growing on them.

Ecological impact of farming

Farming practices simplify food webs, reducing competition by other organisms for human food.

A **monoculture** is where the same crop plant is grown over large areas of land. Crop plants have characteristics such as high yield, disease resistance and high profitability.

Characteristics of a monoculture	Impact on the environment
Crop plant is almost the only plant species present over large areas	Only a few species of herbivore can feed on them, so few food chains supported
Crop plant replaces native plant species	Removes producers for many food chains/webs
Creation of very uniform fields over large areas	The range of habitats and niches for other organisms is reduced
Only crop plants grow in large numbers over large areas	The diversity of the ecosystem is reduced, with fewer species present and fewer individuals of those still present
Large amounts of inorganic fertiliser needed	Damages soil structure and soil organisms, including those involved in recycling of nutrients
Need large amounts of pesticides	Destroys many non-pest organisms and leads to bioaccumulation of pesticides along food chains

Removal of hedgerows gives more land for growing crops and allows the use of larger, more efficient machinery. It also removes 'reservoirs' of pests.

The impacts on the environment are:

- loss of plant and animal species of hedge/woodland communities;
- loss of habitats and niches for woodland community species;
- more extreme abiotic factors, due to removal of shelter;
- removal of 'green highways', which allow plants and animals to spread/migrate.

Organic effluent

This is untreated faeces and urine in liquid form, from sewage works and farms. It may be released into streams/rivers/lakes/sea intentionally or accidentally.

- This provides food for exponential growth of bacteria, increasing oxygen use by bacteria in respiration.
- The Biochemical Oxygen Demand (oxygen used by organisms) increases and the oxygen content of the water falls.
- Abiotic factors in the environment/ecosystem become more extreme.
- Species needing oxygenated water move away/die, including many invertebrates and most species of fish – species that eat fish decrease.
- The diversity of the ecosystem decreases.
- Species adapted to anaerobic conditions colonise the water, forming a community with a smaller food web, with fewer food chains.

Ⓢ Disease resistant crops can be produced by genetic engineering and cloning techniques.

❶ You need to know at least three of the effects of monocultures and removal of hedges. Your answers have to use the terms used in these notes – not vague, general statements that any non-biologist could make.

✓ *Quick check 1, 2*

Nitrates and phosphates

These ions are limiting factors on growth of microscopic and filamentous algae.

- Excess nitrate and phosphate produces eutrophication and can cause exponential growth of algae in water.

- Eutrophication happens when the mineral ions/nutrients in water increase above normal levels, removing limiting factors on the exponential growth of organisms such as algae.

- The algae cover other species of water plants in the community, preventing them from getting enough light and killing them.

- This removes producers in food chains and webs, and reduces oxygen in the water.

- Exponential growth of algae produces more dead algae, providing food for growth of decomposing bacteria and a higher Biochemical Oxygen Demand.

- Decomposition of organic effluent releases nitrate and phosphate ions.

- Excessive use of nitrate/phosphate based fertilisers leads to the ions being washed (not leached) into streams/rivers/lakes after rain.

> Algae **do not** use up the oxygen in water.

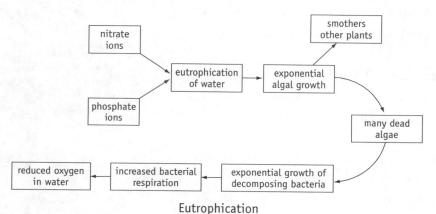

Eutrophication

> ✓ *Quick check 3, 4*

Quick check questions

1. Explain why monocultures are thought to damage ecosystems.

2. Suggest why the government provides grants to farmers to replant hedges.

3. Explain the differences in the effects on freshwater ecosystems of organic effluent and excessive nitrate ions from inorganic fertilisers.

4. Untreated human sewage used to be released into many rivers. This caused fish-eating otters (carnivorous mammals) to vanish from the rivers. Sewage is now treated before being released and otters are gradually returning to the rivers. Suggest how treating the sewage resulted in the return of the otters.

Pesticide toxicity, food production and conservation

Pests are organisms that damage or destroy crops or livestock. There are different types of **chemical pesticide**, for killing different types of pest:

- **insecticides** kill insect pests;
- **herbicides** kill weeds;
- **fungicides** kill disease-causing fungi.

Biodegradable pesticides are broken down by the activities of organisms and only remain in the environment for a short time after they are applied.

Non-biodegradable pesticides cannot be broken down by organisms and remain/persist in the environment for long periods.

Bioaccumulation of non-biodegradable pesticides happens in food chains and webs.

- Pesticide in/on plants (primary producers) is eaten by herbivores (primary consumers) as they feed.
- Herbivores cannot break down the pesticide and it accumulates in their tissues.
- Herbivores may also have pesticide on them as the result of spraying.
- Carnivores (secondary consumers) eat large numbers of herbivores and accumulated pesticide.
- Pesticide accumulates in the tissues of the carnivores – at higher concentrations than in the herbivores, because of the numbers of herbivores eaten.
- This process continues up food chains to the top carnivores – which may be humans.
- At higher trophic levels accumulated pesticide may be concentrated enough to cause death, or serious illness.

pyramid of biomass for a foodchain   DDT concentration

Bioaccumulation of pesticide (ppm, parts per million)

(S) Along with the variation seen in any population, some pest organisms may have resistance to a pesticide – due to a chance mutation. They will have a selective advantage and be more likely to survive and breed when pesticides are used. This leads to pesticide resistant strains.

Pesticides may get into streams, rivers or lakes but they do not cause eutrophication!

Avoid waffle! Do not give the sort of answers expected of someone who has not studied A level Biology; for example, talking about 'pesticide getting into birds of prey', as the answer to a question about the problems of bioaccumulation of pesticide. The appropriate level of response is shown in the notes.

✓ *Quick check 1*

Balancing food production and conservation

Some modern farming practices are damaging the environment. **Destruction of habitats** reduces the diversity of ecosystems, making them less stable and farming **less sustainable**. Balanced judgements need to be made between the need to increase food production and the need to conserve the environment. This may involve the use of alternative strategies.

- Heavy use of **inorganic fertilisers**:
  - does not add humus to soil;
  - damages communities of microorganisms involved in decomposition and the nitrogen cycle;
  - damages soil structure, leading to increased erosion;
  - increases the risk of eutrophication, due to run-off.
- **Organic fertilisers**, such as manure, avoid these problems and release nutrients gradually, as they are decomposed.
- **Erosion** results from:
  - monocultures, which leave fields without plant cover after ploughing, and allows run-off from rain to carry away soil;
  - large fields with few hedges, increasing wind-blow of soil;
  - use of inorganic fertilisers, which destroys soil structure, making it easier to erode.
- These effects can be avoided by:
  - using crop rotation and winter plantings, to keep soil covered;
  - planting hedges as wind-breaks;
  - use of organic fertilisers, to improve soil structure.
- Pesticides should be **biodegradable** and:
  - as pest-specific as possible, to avoid killing other/useful species;
  - used at appropriate times, when a pest outbreak is forecast;
  - used when climatic factors are appropriate, e.g. low wind, to avoid spray drift;
  - applied appropriately, e.g. spot spraying, rather than aerial spraying;
  - alternatives should be considered, e.g. biological control.
- **Habitat variety** can be maintained by:
  - planting hedges and trees in the corners of fields;
  - leaving 'headlands' – strips around the edges of fields unsprayed with fertiliser or pesticide, where wild species of plants and animals grow with the crop.

▶ As before, use the sort of terms used in the notes when answering a question.

✓ *Quick check 2*

? *Quick check questions*

1 Explain *three* reasons for using biodegradable pesticides, rather than non-biodegradable ones.
2 Explain three changes that farmers could make to their farming practices to make them less damaging to the environment. Suggest in each case why they might be reluctant to make the change.

Module 5: end-of-module questions

1 The diagram shows some of the processes in the nitrogen cycle.

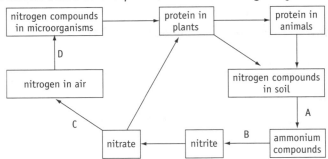

a Name the process involving microorganisms occurring at
stage A, B, C and D. [4]

b Describe how the excessive use of nitrate fertilisers on farmlands
can affect nearby aquatic ecosystems. [6]

c Describe the process involved by which a carbon atom in a glucose
molecule in a mammal could eventually form part of the cell wall
of a plant. [3]

2 The diagram shows the energy flow through an ecosystem.

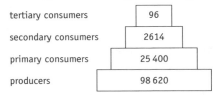

a Suggest units which might be used to record the energy flow between
trophic levels. [2]

b Calculate the percentage energy transfer from producer to primary
consumer. [2]

c Give three reasons why the percentage of energy transferred between
consumers is generally low. [3]

d The diagram shows a pyramid of numbers for a food chain.

i Suggest two reasons why there are more secondary consumers than
primary consumers in this pyramid. [2]

ii Draw a pyramid of biomass for this food chain. [1]

3 A woodland ecosystem will contain populations of differing species, which form one or more communities. Within a woodland each species will occupy an ecological niche and the size of its population will vary as a result of biotic and biotic factors

   a What is meant by: (i) a population; (ii) an ecological niche? [2]

   b Suggest two abiotic factors which could affect the population size of wood violets in woodland. [2]

   c In an investigation into the leaf area of wood violets in two different parts of a woodland, the standard deviations of the two populations were calculated. The results showed that there was a statistically significant difference between the two populations. Explain what is meant by: (i) standard deviation; (ii) a significant difference. [2]

   d Describe the techniques that could be used to compare the vegetation in two areas of woodland. [6]

Appendix: Exam Tips

Lots of marks are lost by not answering questions as they are set, or not knowing material in the syllabus. You must know the information, terms and examples that are included in the syllabus – the exam board only allows questions that can be answered using syllabus material. Other information will not harm you, but will not be necessary to get a good mark!

Describe

'Describe' means put information into words. The information is usually **given** to you in a table, graph or diagram.

Example: the graph shows the rate of reaction of an enzyme with different concentrations of substrate and how the rate is affected by the addition of a particular concentration of an inhibitor.

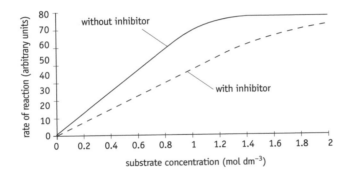

Question

Describe the effect of each of the following on the rate of reaction of the enzyme:

a the concentration of substrate;

b the inhibitor.

Answers

a Between substrate concentrations of 0 and 1 mol dm^{-3}, the rate is directly proportional to the concentration of substrate. The rate reaches a maximum at a substrate concentration of about 1.4 mol dm^{-3}.

b The inhibitor reduces the rate of action of the enzyme at lower concentrations of substrate. Above substrate concentrations of 1.2 mol dm^{-3}, the rate with the inhibitor gets closer and closer to the rate without the inhibitor.

These answers **describe** what you can see on the graph – in reasonable detail.

Explain

'Explain' means that you should 'know' the answer from syllabus material that you have been taught. The following question uses the same graph and information as for 'Describe'.

Question

Explain if the inhibitor is competitive or non-competitive.

Answer

The inhibitor is competitive, because its effect is overcome by increasing the concentration of substrate.

A non-competitive inhibitor would inhibit the enzyme at any concentration of substrate.

Suggest

'Suggest' means that you are unlikely to have been taught about the material in the question but you should have been taught things from the syllabus that will allow you to answer. The following question applies to an enzyme and its inhibitor.

Question

Two substances, **X** and **Y**, were investigated as possible rat poisons. Both inhibit the same enzyme in an important biological process. The diagrams show the structure of the enzyme, its substrate and the inhibitors. Use the information in the diagrams to suggest which inhibitor would be the best poison.

Answer

Inhibitor Y would be best, because it is non-competitive. It binds to a site other than the active site.

Its effect cannot be overcome by more substrate (unlike inhibitor Y) and so the metabolic pathway is blocked.

OR

Inhibitor X would be best, because it is competitive. It binds to the active site. This reduces/stops the substrate being turned into product (and stops the metabolic pathway).

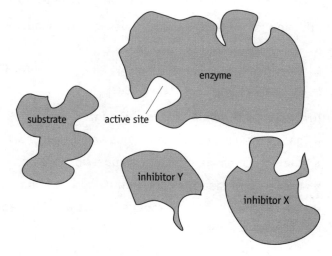

It is not unusual to have alternative answers to this sort of question. You are not expected to have learnt this material – it isn't specifically given in the syllabus. Either answer is a reasonable interpretation of the information.

Synoptic questions

These are part of the exams for Module 5(a), Environment and Modules 6, 7 and 8. Synoptic questions require you to bring together knowledge and ideas from all the previous Modules – AS Modules 1, 2 and 3 and A2 Modules 4 and 5. You are **not expected to remember all the detailed knowledge from previous modules** and you will not be asked for specific details. The synoptic questions will be built into longer questions related to the Module being tested.

Example

In a Module 5 exam you might be asked a series of questions about the roles of saprophytic bacteria and fungi in decomposition and the recycling of nutrients.

- You might then be given data showing that the rate of recycling of nutrients increases in spring and summer, compared to winter and be asked to suggest why this occurs.

- The answer involves warmer climatic conditions in spring and summer, leading to faster rates of action of enzymes of bacteria and fungi and their faster digestion of organic remains. This is using basic knowledge about a factor affecting enzyme activity – from Module 1.

Essays

In the **Biological Principles** exam you will have to write **one** essay. There will be a **choice from two titles**. These will ask about important ideas that apply to many areas of the specification. One example used by the exam board in training material is given below.

The process of diffusion and its importance in living organisms.

Diffusion is involved in many topics in many modules.

- The process itself is in AS Module 1 (AS 1).
- Together with osmosis, as a special case of diffusion of water and how substances enter and leave cells.

Diffusion or osmosis are involved in:

- gaseous exchange in lungs, gills and leaves (AS 1);
- the uptake of the products of digestion (AS 1);
- exchange of materials between blood in capillaries and tissues (AS 2);
- the uptake of water by roots and root pressure, and in the mass flow hypothesis of translocation of sugars (AS 2);
- regulation of blood water potential (A2 4);
- action potentials (A2 4);
- synaptic transmission (A2 4).

This list would make a good **essay plan**!

To get a good mark you need to write about as many of these as you can. You need to include **important concepts** and the **correct terms**. **Do not** write at length about just one example. **Do not** write just about humans – always look for the chance to mention other organisms – especially plants!

Very few good essays will be more than **about three pages** long – there are no marks for the amount you write! You do not need an introduction and conclusion! You must write in continuous prose – no lists or bullet points.

Marks are awarded as follows:

- **16 for knowledge and understanding;**
- **3 for breadth of knowledge;**
- **3 for relevance;**
- **3 for quality of written communication.**

In our example about diffusion, if you write a paragraph about each of the things on the list, you could get all the marks for knowledge and understanding, breadth and relevance.

Answers to quick check questions

Module 4: Energy, Control and Continuity

Energy supply – ATP, oxidation and reduction

1   All organisms respire; plants also photosynthesise.

2   ATP phosphorylates other molecules; making them more reactive; and lowering activation energy needed.

Photosynthesis

1   Photolysis of water; gives electrons to replace those lost from chlorophyll; and hydrogen ions which reduce NAD to NADPH.

2   Light energy absorbed by chlorophyll molecules; excited electrons leave chlorophyll; electrons are a source of chemical energy/reducing power.

3   Hydrogen carbonate ions; glycerate 3-phosphate; ribulose bisphosphate; and glucose.

4   No light, no light dependent stage; NADPH production stops; so soon no NADPH to reduce glycerate 3-phosphate to sugar.

Respiration

1   (6-carbon) glucose oxidised to 2 molecules of (3-carbon) pyruvate; by removal of hydrogen; accepted by NAD to give reduced coenzyme.

2   In Krebs cycle, 6-carbon compound oxidised to a 4-carbon compound; by removal of hydrogen; this is accepted by NAD and FAD to give reduced coenzymes; supply chemical energy/reducing power for ATP production by oxidative phosphorylation.

3   Reduced coenzyme NADH/FAD reduces first protein carrier; electron passes down carrier chain; in series of oxidation and reduction reactions; loses energy as it does; some of the energy used to make ATP.

4   Oxygen is the final electron acceptor from oxidative phosphorylation; if electrons not accepted, all protein carriers soon in reduced state and no electrons flow; so no oxidative phosphorylation or ATP; glycolysis does not produce enough ATP to keep us alive.

Survival and coordination

1   Receptors in eye react to stimulus of light from 'frightener'; send nerve impulses along optic nerve to brain (visual cortex); brain (association area) identifies threat from memory; brain formulates a response; nerve impulses to effectors – adrenal glands release adrenaline – the response.

2   Pain receptors in skin detect heat stimulus; nerve impulses along sensory neurone to spinal cord; information crosses synapse to association neurone; this synapses with motor neurones and neurones carrying information up spinal cord to brain; nerve impulses travel to biceps muscle in arm and the brain; nerve impulses reach biceps first, leading to response – before brain is aware.

3   The hormone is distributed all around the body in the blood, so reaches all cells; any cell with a specific receptor for the hormone has its physiology affected; all the cell/tissue types affected must have a specific adrenaline receptor.

Regulation of blood sugar

1   Insulin levels rise when blood sugar above normal – fall when sugar below normal; sugar levels rise after meals – followed by insulin; sugar levels fall when fasting/exercising – followed by insulin.

2   High blood sugar causes pancreas (islets of Langerhans) to secrete insulin into blood; binds to specific membrane receptors on liver cells; causes glucose channels to open; glucose enters liver cells, lowering blood sugar; provides more substrate for enzymes making glycogen.

3   Liver cells have different specific membrane receptors for insulin and glucagon; insulin binding causes glucose channels to open; glucose enters cells, giving more substrate for enzymes making glycogen; glucagon binding activates phosphorylase enzymes inside the cells that break down glycogen to glucose.

Regulation of body temperature

1   During exercise muscles respire more and produce surplus heat; blood carries heat away; warming of blood detected by thermoreceptors; in hypothalamus; send nerve impulses to heat loss centre in hypothalamus; response coordinated; nerve impulses sent to effectors; responses increase heat loss; circular muscles around arterioles relax → vasodilation; sweat glands → more sweat; muscles at base of hairs relax → hairs lie flat.

2   Cooling blood near stomach causes slight fall in temperature of blood as a whole; this fall detected by receptors in hypothalamus; response is reduction in heat loss from body.

3   Lower body temperature is below optimum for body's enzymes; all enzyme reactions slow, including respiration; less respiration means less heat from respiration; this is what keeps body temperature above environmental; less heat produced than lost, so body temperature continues to fall.

Removal of metabolic waste

1   Body cannot store surplus amino acids; they are deaminated in liver cells; to organic acids used in respiration; and ammonia which enters the ornithine cycle; urea is formed, which is less toxic than ammonia.

2   Filtrate does not contain blood cells; or blood proteins; does contain dissolved urea, glucose and mineral ions (like plasma); because these substances are small enough to pass through by ultrafiltration; in the glomerulus and Bowman's capsule; blood cells and proteins are too large.

3 Useful substances reabsorbed along tubule; glucose, Na$^+$ and K$^+$ in proximal convoluted tubule; by active uptake; by specific carrier proteins; water reabsorbed by osmosis; in proximal and distal tubule, loop of Henle and collecting duct; urea not reabsorbed.

4 If the filter in the glomerulus and Bowman's capsule is damaged; so there are holes large enough for glucose/proteins to cross; very high levels of glucose in the blood (as in untreated diabetes), so too much glucose in filtrate for all to be reabsorbed.

Regulation of blood water potential

1 Active transport of chloride ions by cells of loop of Henle; followed by Na$^+$ ions; gives more negative/lower water potential to surrounding tissues; these also surround collecting duct; water potential lower than filtrate in duct; so water leaves filtrate by osmosis.

2 Water potential of blood becomes slightly lower/more negative; this stimulus detected by receptors in blood vessels in hypothalamus; response is release of ADH from pituitary gland; ADH makes cells of collecting duct and distal tubule more permeable to water; more water is reabsorbed from filtrate; smaller volume of (more concentrated) filtrate, so less water loss.

The eye

1 Accommodation/fine-focusing; ciliary muscles contract; taking tension out of suspensory ligaments; elasticity of lens makes it more biconvex/fatter; shortens focal length of lens/keeps image of print in focus on retina.

2 Little light at night, so only rods stimulated; outside of the fovea, in edge of vision; when looking straight at, light falls on fovea; only cones there and they do not react to dim light.

3 Light bleaches the pigment rhodopsin in membranes of vesicles; rhodopsin breaks down to retinene and scotopsin; alters permeability of membrane to Na$^+$ ions; nerve impulses formed that pass along optic nerve; carrying information.

Colour vision

1

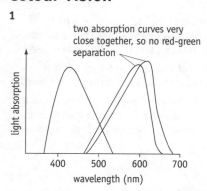

two absorption curves very close together, so no red-green separation

light absorption

wavelength (nm)

2 Light focused on fovea where cones are; cones give colour vision; each cone synapses with one bipolar and one ganglion cell; so sends information along optic nerve to brain about its very small area of the retina; gives high visual acuity.

Nerve impulse

1 Long axon, carries nerve impulses long distances; axon myelinated – nerve impulses 'jump' from one node of Ranvier to the next → fast transmission; many dendrites, to synapse with other neurones.

2 Travelling depolarisation; transmembrane protein channels open for Na$^+$; they diffuse in along their diffusion gradient; resting potential across membrane falls locally – depolarisation; inrush of sodium ions makes potassium channels open; they diffuse out along diffusion gradient; start to restore membrane potential; sodium channels close and sodium pump pumps out sodium ions again; the local depolarisation opens sodium channels in next section of axon, so depolarisation moves along.

3 Refractory period 2 milliseconds; so 500 nerve impulses per second.

4 Small stimuli below threshold for initial depolarisation of cell membrane, so no action potentials; information on size of stimulus carried as the frequency of nerve impulses; the larger the stimulus, the greater the frequency of nerve impulses; refractory period means limit to frequency of nerve impulses and size of stimulus we can detect.

Synapses and drugs

1 Action potential reaches presynaptic membrane; depolarisation opens calcium channels – calcium ions diffuse in; vesicles with acetylcholine fuse with membrane; acetylcholine released and diffuses across narrow synaptic cleft; binds to specific receptors on postsynaptic membrane; makes Na$^+$ channels open; leads to depolarisation and if enough of these, an action potential.

2 Information carried across synapse by bursts of acetylcholine, corresponding to action potentials at presynatic membrane; this produces distinct depolarisations in postsynaptic membrane; if transmitter remained in synaptic cleft, constant depolarisations and no information.

3 Synapse between two neurones; neuromuscular junction between neurone and muscle fibre; depolarisation of postsynaptic membrane leads to action potentials; depolarisation of motor end plate membrane leads to entry of calcium ions and contraction.

4 Venom molecule has part with a shape that fits acetylcholine receptor; competes with acetylcholine and occupies binding sites; does not cause depolarisation of postsynaptic membrane; but stops acetylcholine, so no action potentials and paralysis.

Brain and cerebral hemispheres

1 Sensory area/visual cortex receives nerve impulses from the retina/eye; association area interprets information/nerve impulses from sensory area; compares against memory to identify friend.

2 Visual cortex at the back of the head; injury to this area could affect sight; if left eye faulty, then right side of brain affected/vice versa.

3 Auditory motor area; since not able to form speech; not association area, since able to understand spoken instructions.

Autonomic nervous system

1 *Two* of: iris dilates, allowing more light into eye to stimulate receptors – to see threat clearly; bronchi dilate, allowing easier movement of gases in and out of lungs – improving ventilation of lungs/oxygen uptake; increase in heart rate, increasing blood flow to muscles/lungs – so more oxygen to muscles for respiration; vasoconstriction of blood vessels, raising blood pressure/reducing flow to non-essential tissues – so better blood supply to muscles/essential organs; reduced blood supply to gut, so more blood to flow to muscles – more oxygen/glucose for respiration.

2   Parasympathetic reflex; too much light (stimulus) on receptors in retina; nerve impulses to parasympathetic nervous system; coordinates response; sends nerve impulses to effectors; circular muscles in iris contract; response is a reduction in diameter of pupil and less light entering eye.

Muscle structure and function

1   Biceps contracts; triceps relaxes; radius pulled up.

2   Myosin and actin.

3   The H zone.

Muscle contraction

1   i   Remains the same.

    ii   Decreases.

2   Hydrolysis of ATP, to release myosin head from actin.

3   Bind to troponin, causing tropomyosin to move from the binding site.

Genotype and meiosis

1   i   The genes an organism has and the alleles of those genes.

    ii   The characteristics of an organism due to its genotype and interactions between the genotype and environment.

2   A and B alleles both dominant to O allele; so types A and B can be heterozygous carriers of O; this gives 50% chance of a gamete containing the O allele; since meiosis produces gametes with one allele of each gene.

3   Meiosis produces haploid cells with one copy of each chromosome; when gametes fuse, the diploid number of chromosomes is restored; if diploid gametes fused, the number of chromosomes would double with each generation.

Sex determination, monohybrid inheritance and sex-linkage

1   Father determines sex of offspring; 50% of sperm carry X and 50% Y; so 50% chance with each pregnancy of a girl.

2   Since all F_1 round, round is dominant to wrinkled; all F_1 heterozygous; expect 25% of F_2 to be homozygous and wrinkled; 3:1 ratio round to wrinkled.

3   If neither parent affected, then mother carries haemophilia; recessive allele on X chromosome; 50% of her eggs have haemophilia allele; so 50% of sons affected (since they get a Y from father); 50% of daughters will be carriers. *You could use diagrams to help your explanation.*

Dihybrid crosses

1   Since F_1 all purple flowers and green pods, purple dominant to white and green dominant to yellow; F_1 heterozygous PpGg; gametes PG, Pg, pG, pg in equal numbers; Punnett square should give 9:3:3:1 ratio; 9 purple, green : 3 purple, yellow : 3 white, green : 1 white, yellow.

2   Epistasis; if recessive allele of gene produces a non-functional enzyme; then homozygous recessive blocks metabolic pathway; other genes produce functional enzymes but their products not used, or no substrate available. *You could use diagrams to help your explanation.*

Variation

1   IQ and length of hair are continuous; there is a range between two extremes; it can be measured; there are no distinct groups. Haemophilia is discontinuous; it cannot be measured; there are two distinct groups, those with it and those without.

2   Blood groups are genetically determined; with little/no environmental factors; so, might easily have different blood groups for non-identical, since they have different genotypes; for height and weight, environmental factors important; growing in same environment these are likely to be more similar.

3   Crossing over at chiasma; produces new combinations of maternal and paternal alleles; independent assortment; produces new combinations of maternal and paternal chromosomes and their alleles.

Natural selection

1   Variation in population; some phenotypes better adapted to environment than others; compete better, have a selective advantage; produces differential survival rates; better adapted; more likely to survive long enough to reproduce.

2   Using warfarin changes the environment of rat populations; new selection pressure; by chance, some rats had a mutation giving resistance to warfarin; gives higher differential survival rate; more likely to survive and pass on resistance alleles/genes; produces increasing frequency of resistance alleles in population.

3   Answer implies that the rats decided to develop a resistance gene/allele in response to warfarin; if an allele/gene giving resistance is present, it is due to a chance mutation.

Speciation

1   Populations geographically isolated from those on other islands; environment on each island different to other islands; different selection pressures; natural selection produced changes in allele/gene and phenotypic frequencies; until populations became reproductively isolated from each other; unable to breed with them to produce fertile offspring.

2   a   Reduction in number of digits/toes to one; bones of this one elongated; runs on tip of one toe; reduction in length of humerus; radius and ulna shorter and fused.

    b   Increase in length of legs; longer legs for faster running; greater height allows predators to be seen.

3   Evolution is a process, it has no aim; it is a change in the gene frequencies of a population; in a stable environment this tends to favour 'average' phenotypes and keep a species the same.

Classification

1   Should be written as Homo sapiens.

2   Humans and chimpanzees are both primates; being in the same order, they share a common ancestor; which had the genes they share.

3   Plants photosynthetic, fungi saprophytic or parasitic; plant cell walls of cellulose, fungi of chitin; plants multicellular, fungi unicellular or hyphae.

Module 5: Environment

Energy flow through ecosystems

1 Very little energy passes from one trophic level to the next; by 5/6 trophic levels there is no energy to pass on/not enough to support another level.

2 $11 \div 15\,000 \times 100 = 0.07$.

3 Pyramid of number – the number of individuals in a given area at each trophic level at a specific time. Pyramid of mass – mass (kg) of organisms at each trophic level at a specific time and in a given area (m^2) – units of kg/m^2. Pyramid of energy – energy (kJ) in each trophic level, for a particular area (m^2) and period of time (y) – units of $kJ/m^2/y$.

4 Difficult to find energy content of organisms; would destroy organisms; hard to know how much of the energy present was for that year.

Nutrient cycles

1 Photosynthesis is where inorganic carbon in CO_2 enters the biotic environment; respiration is where carbon returns to the abiotic environment as CO_2.

2 Decomposers such as bacteria and fungi; feed on dead organisms and respire, returning carbon to the abiotic environment; they also excrete ammonia from amino acid metabolism; nitrogen-fixing bacteria convert nitrogen into nitrate ions plants can use; to make amino acids/proteins; Rhizobium fixes nitrogen in root nodules of leguminous plants; nitrifying bacteria convert ammonia to nitrites and nitrates; denitrifying bacteria break down nitrates, releasing nitrogen.

3 a Nitrogen fixation by nitrogen-fixing bacteria/Rhizobium; into ammonia/nitrites/nitrates plants can use; nitrifying bacteria convert ammonia to nitrites and nitrates that most plants prefer; plants use ammonia/nitrites/nitrates to synthesise amino acids and then proteins.

  b There is usually more than enough carbon dioxide and water for photosynthesis; in growing seasons the temperature is not a limiting factor; nitrates are needed for protein synthesis; including enzymes for metabolism; proteins for cell growth; protein in seeds/grain/fruit; soil water contains few nitrate ions; which are easily leached/highly soluble; are removed with the harvest.

Ecological techniques

1 Line transect from field into wood; use frame quadrat divided into 100 squares at regular intervals along the transect – say every 5 m; record % cover for each species under the quadrat; by counting the number of squares covered by each species; repeat many times along different transect lines, to avoid one unusual set of results.

2 Select one natural site where oak trees grow (not planted); and one where another tree/plant dominates and oaks do not grow; record abiotic factors at regular intervals each year (say every month); record light with a light meter; air and soil temperatures with a thermometer; soil pH with universal indicator; average results for each recording time after several years; to avoid single inaccurate/anomalous results; compare results from the two sites, to identify which factor (if any) is more important.

Ecosystems

1 E.g. hands to make/use tools; forward facing eyes, to judge distances/stereoscopic vision; ears on side of head, to judge direction of sounds; stand upright, to see predators coming; large cerebral hemispheres, for intelligent behaviour manipulating our environment; endotherm/warm-blooded, to survive and function in cool environmental conditions.

2 Availability of mineral ion, e.g. nitrate; becomes limiting factor; due to intra- and interspecific competition for nitrate; so population growth stops and there is a decline in numbers.

3 Interspecific competition between prey organism and new organism; leading to decline in numbers of prey organism; carrying capacity falls; followed later by fall in predator numbers.

Succession and climax communities

1 Changes in communities over time with one following another, until climax community established; often starts with colonisers on bare soil; this community makes biotic and abiotic factors less severe; especially soil factors; e.g. humus to hold water; making it possible for other plants to establish a new community; which outcompete original colonisers; repeated with later communities, until climax community – stable over long period of time.

2 Herbaceous plants change soil factors; when they die and decompose; add humus to soil which holds water; source of mineral ions such as nitrate; improve structure of soil; plants also act as wind breaks, increasing temperature locally/reducing water loss; make abiotic conditions less severe, so seedlings of woody plants can grow.

3 Interspecific competition; marram cannot compete with woodland plants; seedlings cannot get established.

Ecological impact of farming

1 Few herbivores feed on them, so few food chains/simpler food webs; replace other species – remove producers for food chains; reduced habitats and niches; lower diversity, numbers of organisms and species; lots of fertiliser used, damages soil; lots of pesticides used, destroy non-pest organisms; can bioaccumulate and harm higher trophic levels of food chains/webs.

2 Creates green highways – organisms can spread/migrate; more habitats/niches for woodland organisms; preserves woodland species/communities.

3 Organic effluent provides food for bacteria; reproduce exponentially; bacterial respiration increases and reduces oxygen in water; oxygen using species die/move away, such as fish; nitrate ions produce eutrophication; allows exponential growth of algae – smothers other vegetation; dies and provides lots of food for decomposing bacteria; their exponential growth lowers oxygen in water.

4 Treated sewage does not support large bacterial populations; or the anaerobic organisms associated with them; other organisms re-colonise the water, including fish; more fish means food for otters, so they move into these areas.

Pesticide toxicity, food production and conservation

1 Biodegradable pesticides broken down by organisms; no bioaccumulation; carnivores at top of food chains not killed/harmed; not concentrated in human food.

2 E.g. use organic fertilisers: adds humus to soil; food for decomposers in nutrient/nitrogen cycle; less risk of eutrophication. Harder to handle – spray inorganics. Plant hedges: creates green highways – organisms can spread/migrate; more habitats/niches for woodland organisms; preserves woodland species/communities. Use land that could grow crops and make money. Avoid monocultures: more food chains supported; greater biodiversity; more habitats and niches. Best market for monoculture crops, so less income.

Answers to end-of-module questions

Module 4: Energy, Control and Continuity

1 a i ATP, reduced NADP, water.

ii Grana.

b i Five.

ii Three.

c Reduced NADP; provides hydrogen for reduction.

d Produced from glycerate 3-phosphate.

2 a i Cytoplasm.

ii Mitochondrion.

b i Three.

ii Four.

c Hydrogen/electrons pass down carriers. Carriers are at decreasing energy levels. Energy released used to phosphorylate ADP to ATP.

3 a Maintenance of a constant internal environment.

b Pancreas detects rise or fall in concentration. Rise results in release of insulin. Fall results in release of glucagon. Hormones travel to liver in blood. Hormones stimulate enzyme-controlled reactions. Insulin stimulates conversion of glucose to glycogen. Glucagon stimulates conversion of glycogen/lipids/protein to glucose. Insulin increases permeability of cells to glucose/conversion of glucose to lipid. Reference to negative feedback mechanism.

c Removal of amine group and hydrogen from amino acids. Reference to ornithine cycle. Production of urea.

4 a i Blue.

ii Orange.

b White.

c Several rod cells connect to a single bipolar neurone. Summation at low intensity light stimulation generates an action potential.

d Circular ciliary muscles contract. Suspensory ligaments slacken. Lens becomes more convex. Due to its own elasticity. Converging power is increased.

5 a Cation pump. Membrane of neurone differentially permeable. Exit of sodium ions greater than entry of potassium ions.

b Stimulus must be above threshold intensity before an action potential is set up. Amplitude of impulse remains constant.

c Entry of calcium ions into synaptic knob. Vesicles fuse with presynaptic membrane. Release of neurotransmitter. Diffuses across synaptic cleft. Attaches to receptors on postsynaptic membrane. Action potential/end plate potential in muscle fibre.

6 a Reference to cerebral hemispheres. Impulses from receptors to sensory area. Interpretation by association centre. Impulses from motor area to effectors/muscles.

b Impulses from photoreceptors via sensory neurones to brain. Automatic/involuntary/autonomic control. Impulses via sympathetic neurones to muscles of iris. Stimulates contraction of radial muscles and relaxation of circular muscles.

7 a A = myosin; B = actin.

b i Shortens.

ii Remains the same.

c Formation of actin–myosin cross-bridges. ATP required to form/break bridges. Actin filaments pulled over myosin filaments. Shortening of sarcomere. Ratchet mechanism. Calcium ions activate contraction. Correct reference to role of tropomyosin.

8 a Four chromosomes, one from each homologous pair.

b Crossing over/chiasmata. Independent assortment of chromosomes.

9 a i A different form of a gene.

ii I^0.

iii I^A and I^B.

b

parental phenotypes	blood group A	blood group B
parental genotypes	$I^A I^o$	$I^B I^o$
genotypes of gametes	$I^A I^o$	$I^B I^o$
genotypes of children	$I^A I^B$, $I^A I^o$, $I^B I^o$, $I^o I^o$	
phenotypes of children	AB, A, B, O	

10 a Organisms which breed to produce fertile offspring.

b Isolation; no gene flow between populations; variation in population; due to mutations; natural selection; change in allele frequency; change occurs over a long period of time.

c Hierarchy – groups within groups; no overlap; similar characteristics; reflecting evolutionary history, binomial nomenclature.

Module 5: Environment

1 a A = ammonification/decomposition; B = nitrification; C = denitrification; D = nitrogen fixation.

b Phosphates/nitrates run off into water; increase in plant/algae growth; reduced light to submerged plants results in death; increase in decomposers/bacteria; use oxygen in respiration; decrease in aerobic organisms; low species diversity; increase in anaerobic organisms.

c Respiration produces carbon dioxide; photosynthesis uses carbon dioxide to form glucose; condensation of glucose molecules to form cellulose.

2 a Energy unit per unit area per unit time period, e.g. $kJ/m^{-2}/y^{-1}$.

b 25.75%.

c Heat/respiration; movement/muscle contraction; indigestible material/excretion.

d i Secondary consumers are parasites; rapid reproductive cycle; small size.

ii

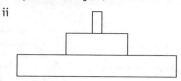

3 a i A group of organisms of the same species living in the same habitat and able to interbreed.

ii Reference to feeding role; habitat role.

b Light intensity, soil pH, nutrient availability, water content of soil.

c i Degree/spread of variation from the mean.

ii Rejects the null hypothesis/difference is due to factors other than chance.

d Quadrats; large sample; randomisation; counts/percentage cover; diversity index; point quadrat; belt transect; line transect; use of averages.

Index